ENERGY POLICIES, POLITICS AND PRICES

EMPLOYMENT EFFECTS OF TRANSITION TO A HYDROGEN ECONOMY IN THE U.S.

ENERGY POLICIES, POLITICS AND PRICES

Nuclear Power's Role in Generating Electricity
Perry G. Furham
2009. ISBN: 978-1-60741-226-7

OPEC, Oil Prices and LNG
Edward R. Pitt and Christopher N. Leung (Editors)
2009. ISBN: 978-1-60692-897-4

OPEC, Oil Prices and LNG
Edward R. Pitt and Christopher N. Leung (Editors)
2009. ISBN: 978-1-60876-614-7 (E-book)

Energy Prices: Supply, Demand or Speculation?
John T. Perry (Editor)
2009. ISBN: 978-1-60741-374-5

Dynamic Noncooperative Game Models for Deregulated Electricity Markets
Jose B. Cruz, Jr. and Xiaohuan Tan
2009. ISBN: 978-1-60741-078-2

Power Plant Characteristics and Costs
Stan Kaplan
2009. ISBN: 978-1-60741-264-9

Hydrogen Fuel Perspectives
Ian S. Rubio (Editor)
2009. ISBN: 978-1-60692-444-0

Natural Gas Markets and Lessons Learned
E. K. Cho (Editor)
2010. ISBN: 978-1-61668-249-1

Natural Gas Markets and Lessons Learned
E. K. Cho (Editor)
2010. ISBN: 978-1-61668-429-7 (E-book)

Strategic Petroleum Reserve
Albert L. Strait (Editor)
2010. ISBN: 978-1-60692-290-3

Solar Energy Technologies: From Research to Deployment
Liam G. White (Editor)
2010. ISBN: 978-1-60741-323-3

Federal Energy Management and Government Efficiency Goals
Amelia R. Williams (Editor)
2010. ISBN: 978-1-60692-985-8

Federal Energy Management and Government Efficiency Goals
Amelia R. Williams (Editor)
2010. ISBN: 978-1-61668-538-6 (E-book)

U.S. Energy: Overview of the Trends, Statistics, Supply and Consumption
Gregor E. Peake (Editor)
2010. ISBN: 978-1-60876-041-1

Rational Use and Energy Planning: A Thermodynamic and Geographical Approach
Giuseppe Grazzini, Carla Balocco and Giovan Battista Andreani
2010. ISBN: 978-1-60741-350-9

World Biofuels Production Potential
Thomas E. Rommer (Editor)
2010. ISBN: 978-1-61668-663-5

World Biofuels Production Potential
Thomas E. Rommer (Editor)
2010. ISBN: 978-1-61668-425-9
(E-book)

Transition to Hydrogen Fuel Cell Vehicles
Selim Koca (Editor)
2010. ISBN: 978-1-60741-806-1

The Role of Auctions in Emission Allowance Allocations for Greenhouse Gases
Aubrey D. O'Connor (Editor)
2010. ISBN: 978-1-60741-699-9

The Role of Auctions in Emission Allowance Allocations for Greenhouse Gases
Aubrey D. O'Connor (Editor)
2010. ISBN: 978-1-61668-652-9
(E-book)

Employment Effects of Transition to a Hydrogen Economy in the U.S.
Michele Auriemma (Editor)
2010. ISBN: 978-1-60741-808-5

Reducing Greenhouse Gas Emissions
Joseph G. Levitt (Editor)
2010. ISBN: 978-1-60741-890-0

Reducing Greenhouse Gas Emissions
Joseph G. Levitt (Editor)
2010. ISBN: 978-1-61668-730-4
(E-book)

Energy Efficiency through Combined Heat and Power or Cogeneration
David H. Thomas (Editor)
2010. ISBN: 978-1-61668-341-2

Energy Efficiency through Combined Heat and Power or Cogeneration
David H. Thomas (Editor)
2010. ISBN: 978-1-61668-432-7
(E-book)

U.S. Energy and the Environment: An Overview and Comparative Analysis
Roland H. Terrison (Editor)
2010. ISBN: 978-1-61668-017-6

U.S. Energy and the Environment: An Overview and Comparative Analysis
Roland H. Terrison (Editor)
2010. ISBN: 978-1-61668-641-3
(E-book)

The Smart Grid and Electric Power Transmission
Caitlin G. Elsworth (Editor)
2010. ISBN: 978-1-61668-223-1

Combined Heat and Power - Analysis of Various Markets
Jordan A. Cory
2010. ISBN: 978-1-60741-269-4

Combined Heat and Power - Analysis of Various Markets
Jordan A. Cory
2010. ISBN: 978-1-61668-377-1
(E-book)

Worldwide Biomass Potential: Technology Characterizations
R. L. Bain
2010. ISBN: 978-1-60741-267-0

The Completion of the Oil Era: The Economic Impact
Carlos A. Rossi
2010. ISBN: 978-1-60741-340-0

Carbon Dioxide Emissions
James P. Mulligan (Editor)
2010. ISBN: 978-1-60692-229-3

Carbon Dioxide Emissions
James P. Mulligan (Editor)
2010. ISBN: 978-1-61728-096-2
(E-book)

Ethanol Economics and Ethanol's Impact on Food Prices and Greenhouse Gas Emissions
Darlene E. Marshall (Editor)
2010. ISBN: 978-1-60876-081-7

Ethanol Economics and Ethanol'sImpact on Food Prices and Greenhouse Gas Emissions
Darlene E. Marshall (Editor)
2010. ISBN: 978-1- 61728-073-3
(E-book)

ENERGY POLICIES, POLITICS AND PRICES

EMPLOYMENT EFFECTS OF TRANSITION TO A HYDROGEN ECONOMY IN THE U.S.

MICHELE AURIEMMA
EDITOR

Nova Science Publishers, Inc.
New York

For permission to use material from this book please contact us:
Telephone 631-231-7269; Fax 631-231-8175
Web Site: http://www.novapublishers.com

LIBRARY OF CONGRESS CATALOGING-IN-PUBLICATION DATA

Employment effects of transition to a hydrogen economy in the U.S. / editor, Michele Auriemma.
p. cm.
Includes index.
ISBN 978-1-60741-808-5 (hbk.)
1. Clean energy industries--Economic aspects--United States. 2. Hydrogen as fuel--Economic aspects--United States. 3. Employment forecasting--United States. 4. Labor market--United States. I. Auriemma, Michele. II. Title: Employment effects of transition to a hydrogen economy in the US.
HD9502.5.C543U646 2009
331.7'65810973--dc22

2009038387

Published by Nova Science Publishers, Inc. ✦ New York

CONTENTS

EXECUTIVE SUMMARY*

E.1. BACKGROUND TO THE STUDY

Section 1820 of the Energy Policy Act of 2005, "Overall Employment in a Hydrogen Economy", requires the Secretary of Energy (Secretary) to carry out a study of the likely effects of a transition to a hydrogen economy on overall employment in the United States. This study, prepared by the Department of Energy (DOE or Department) in response to Section 1820, represents the Department's "best efforts" to predict the employment impacts of such a transition. Prior to publication, the study was reviewed by an independent panel of experts from industry and academia. While the panel provided valuable additional input,[1] any study of potential future impacts necessarily presents several difficult challenges, including the choice of a sound methodology and the identification of reasonable data inputs and factual assumptions, given the uncertainties. In addition, technological process is extremely difficult to predict and economic forecasts are often unreliable even over the short term. The results must be interpreted in light of these unavoidable limitations.

E.2. METHODOLOGY

This study estimated the employment impacts of a transformation of the U.S. economy to the use of hydrogen between 2020 and 2050. This time frame was

---

* This is an edited, reformatted and augmented version of a United States Department of Energy Report to Congress – Effects of a Transition to a Hydrogen Eonomy on Employment in the United States dated July 2008.

selected because the ongoing efforts to develop hydrogen based transportation (or "mobile") and stationary technologies indicate that broad-based commercial and industrial use of and the first significant employment impacts from those technologies are most likely to emerge within the indicated time frame. More specifically, hydrogen production technologies are examined as of three dates-2020, 2035, and 2050-and the share of production contributed by each technology is estimated on the basis of DOE's Hydrogen Analysis (H2A) models.[2]

Because it is not possible to predict with precision the rate at which hydrogen technologies will be incorporated into the nation's economy on the above three dates, the study also considers the differences in employment by industry under two scenarios. The more rapid transformation scenario assumes the success of the President's Hydrogen Fuel Initiative (HFI)[3] of saving 11 million barrels of oil per day by 2040 (the "HFI Scenario"), and the less rapid scenario follows DOE's analysis supporting its 2007 program benefits estimation (the "Less Aggressive Scenario").[4] Under the HFI Scenario, for example, the penetration of the light-duty vehicle stock with hydrogen powered vehicles by 2050 is 96%. Under the Less Aggressive Scenario, such penetration by 2050 is 38%.

The study further estimates national employment impacts from the transition to a hydrogen economy using an established economic impact analysis model called IMPLAN (an acronym for "Impact analysis for PLANning"). IMPLAN has been used in other contexts to derive economic impacts of long-term technological developments.[5] The current study uses the IMPLAN inter-industry model with 509 industrial sectors for the U.S. economy. Revised industry purchase vectors were constructed for three industries for the hydrogen scenarios—vehicle components, vehicle assembly, and hydrogen production. Stationary and portable fuel cells were assumed to be produced in the vehicle components vector together with the automotive fuel cells. The data for these revisions were derived from application of DOE's H2A production and delivery models and separate cost estimates for fuel cell vehicles and stationary fuel cells. Technological process is extremely difficult to predict and economic forecasts are often unreliable even for short term predictions. Though this study represents our best estimate, it is important to appreciate the tremendous uncertainties involved.

E.3. Scenarios Shaping Future Hydrogen Markets

HFI Scenario. The base case against which the employment impacts of hydrogen market expansion are compared is a predominantly gasoline economy.

The HFI Scenario assumes rapid market penetration of hydrogen vehicles. The first sales occur in 2018. By 2020, 27% of new vehicle sales are hydrogen vehicles; by 2035, 89%; and 100% by 2050. This results in stocks of light duty hydrogen vehicles in use, respectively, of 3%, 60%, and 96% of the total of light duty vehicles in use in those years. For hydrogen production, it is assumed that distributed reforming of natural gas provides the greatest share of hydrogen production in 2020. Natural gas feedstocks are then largely superseded by 2035 and replaced by coal gasification with carbon sequestration and, to a lesser extent, by biomass gasification. By 2050, biomass and wind, combined, provide 35% of hydrogen supplies. Hydrogen production from nuclear resources provides small shares in both 2035 and 2050. Hydrogen will also be used to fuel stationary fuel cells. In 2020, stationary fuel cells are assumed to supply 1% of the post-2015 growth in national electricity demand. It is assumed that this will increase to 5% in 2035; and to 10% in 2050.

Less Aggressive Scenario. Market penetration of hydrogen vehicles is slower under the Less Aggressive Scenario, reaching approximately 1% of new sales in 2020, 20% in 2035 and 63% in 2050, resulting in hydrogen vehicle stocks of ½%, 7% and 38%, respectively. The composition of hydrogen production technologies over time is similar to that in the HFI Scenario, but output is lower, in line with the smaller number of hydrogen vehicles. The shares of incremental electricity demand provided by stationary fuel cells are zero in 2020, 2% in 2035, and 5% in 2050.

Table E.1 summarizes the Hydrogen Vehicle Adoption and Stationary Fuel Cell Use in 2050 Under the Two Scenarios.

E.4. EMPLOYMENT CREATION AND REPLACEMENT AT THE NATIONAL LEVEL

By 2050, under the HFI Scenario, the transformational adjustments are fully completed, with no more anticipatory investment; U.S. employment is increased by a net of 0.37%, or 675,000 jobs out of a total projected base-case employment of 184 million. Under the Less Aggressive Scenario, the transformation is not fully completed by 2050 and U.S. employment is increased by a net of 0.20%, or 361,000.

The more significant changes occur under the HFI Scenario. Thus, under the HFI Scenario, net employment in the automotive industry is unchanged between the gasoline and hydrogen economies, but replacement of gasoline-related skills with hydrogen-related skills is substantial. For example, in automotive parts manufacturing, 10,000 white collar jobs are created by 2035, replacing an equal number of outmoded jobs. The figure is 12,000 by 2050. Blue-collar employment job creation and replacement in automotive parts manufacturing is 104,000 by 2035 and 117,000 by 2050. The greatest job creation and replacement is for automobile dealerships and repair: 436,000 by 2035 and 680,000 by 2050.

Training implications of the job changes vary by industry and skill, as indicated by a survey of industry opinions. The changes will be substantial in some cases, though many will be evolutionary, resulting partly from experience with hybrids. The needs for new skills will be spread over a number of years. Much of the obsolescence of skills will be met by normal retirement rates. Most of the needs for new skills can be supplied by normal rates of entry into the labor force as workers receive training in new, hydrogen-related skills.

With the automobile industry taking a small proportion of all engineers in the U.S., engineering schools easily have the capacity in the long run to respond to changes in the fields of engineering that will be required. This is also true for the most part in the shorter run, in view of university expertise currently being developed in hydrogen technology R&D.

Table E.1. Hydrogen Fuel Cell Vehicles and Stationary Fuel Cell Uses in 2050 in the HFI and Less Aggressive Scenarios

Scenario		HFI Scenario	Less Aggressive Scenario
Sales	million light-duty vehicles sold/yr	23.9	15.1
	% of all light-duty vehicles sold	100%	63%
Stock	million light-duty vehicles in use	347.5	144
	% of all light-duty vehicles in use	96.0%	38.2%
Hydrogen fuel use	quads/yr[a]	8.02	3.73
	billion gge/yr[b]	64.1	29.95
Stationary fuel cells	percent of incremental electricity demand after 2015 supplied	10%	5%

[a] 1 quad = 1 quadrillion Btu.

[b] gge = gallons of gasoline equivalent.

Some but not all of the blue collar training needed to switch from gasoline to hydrogen vehicles can be provided on the job and by in-house classes. Automotive technicians and mechanics are a large group requiring much training and re-training. They are the group for which bottlenecks appear to be most likely.

Up to 110,000 technicians and mechanics will need to be equipped with hydrogen-technology skills by 2020 to service new vehicles on the road by that year in the HFI Scenario. By 2035, the number is 335,000. By 2050, the number is 630,000, many of whom will have entered the labor force prior to the beginning of hydrogen market expansion.

E.5. Regional Variations in Economic Impacts

Five regions were selected for examination: 1) the Upper Midwest, consisting of Ohio, Michigan, Indiana, Illinois, and Wisconsin; 2) Lower New England and the Upper Mid-Atlantic, consisting of Massachusetts, Connecticut, Rhode Island, New York, Pennsylvania, and New Jersey; 3) California; 4) Tennessee; and 5) the Houston metropolitan area. Unique characteristics of these regions lead to impacts that differ from those at the national level. The Upper Midwest is America's long-established automobile manufacturing belt. The Lower New England and the Upper Mid-Atlantic regions are very diversified economically. California is also a very diversified state, but its industrial mix has been changing in the past decade. Tennessee represents the new automotive manufacturing region of the South. Houston is highly concentrated in petroleum production and is a net exporter of refined products.

Compared to the base case without a hydrogen market transformation, the HFI Scenario leads to a projected gain in the Upper Midwest of 110,000 additional jobs by 2050. The gains will be distributed across 41 industries. The region will lose 4,800 jobs concentrated in 14 industries. The region's net creation of 105,000 jobs is 0.44% of its 2050 base-case employment. Scientific and technical services employment would grow to support both the technical needs of hydrogen production and the technological changes in the automotive industry, while fabricated metals would lose employment.

The Lower New England and Upper Mid-Atlantic region is projected to increase its 2050 employment by 0.56% over what it would be in an all-gasoline economy. The gains are primarily in the production and delivery of hydrogen, while losses are in the corporate offices of upstream energy companies.

California experiences a 0.45% increase over the base case in the hydrogen market expansion under the HFI Scenario. Its high-tech sectors participate in the development of the new hydrogen technologies, as does carbon and graphite manufacturing. The state suffers some loss of employment in petroleum refining relative to the all-gasoline scenario.

Tennessee gains 0.5% in employment relative to the gasoline scenario under the HFI. The state has no significant losses to dampen the gains in hydrogen production and technological and engineering services.

Houston gains 0.37% in employment in the HFI Scenario relative to the base case. Its refining industry suffers in the hydrogen market expansion, compared to the all-gasoline base case scenario, but this impact could be cushioned by the relative ease of retrofitting the area's refineries. Houston's experience in a variety of energy industries helps it gain employment in hydrogen production and in the design and production of energy and chemical pipeline equipment.

E.6. INTERNATIONAL COMPETITION

The transformation to a Hydrogen Economy will serve at least two major objectives in the international area. First, reduction in oil imports, with the attendant increase in energy independence, is a clear U.S. goal to which hydrogen will contribute. Second, if U.S. companies are able to forge a lead in hydrogen technologies, U.S. global competitiveness will be fostered. The movement to hydrogen in particular could well be an opportunity for U.S. automotive firms to recapture market share lost to foreign multinationals in recent years.

Due to overseas operations of U.S. and multinational corporations, with or without a hydrogen transformation, most vehicle production and employment will continue to be tied to countries with large automobile demands. Some effects on the international location of supplies of particular individual materials could be brought about by a transformation to hydrogen. If the production shares held by U.S. and multinational corporations are affected because some companies get ahead of others in introducing hydrogen vehicles, the location of automobile production within the U.S. could be affected in turn due to the fact that companies differ in their regional concentrations of production capacity. Hydrogen, however, will be produced domestically in either case and will be essentially a non-internationally traded commodity. Inasmuch as natural gas is not projected to be a significant long-term feedstock for hydrogen production, little effect on gas imports is projected.

Oil imports, on the other hand, will fall as gasoline is replaced with hydrogen. By 2050, the difference in oil imports between an all-gasoline economy and a hydrogen economy under the HFI Scenario is projected to be $370 billion per year, or 1% of the $38.12 trillion estimated gross domestic product in 2050. Some of the reduction in expenditures on imported oil will be redirected to a domestically produced, largely nontradable good—hydrogen. The $370 billion reduction in oil imports equals approximately 7.5% of projected total U.S. imports in 2050 if trade grows at the same rate as gross domestic product.

The oil import reduction introduces a trade imbalance that will lead to a combination of adjustments in non-oil imports and exports. The adjustments will be spread over the many non-oil commodities involved in U.S. trade. Considering that both import and export adjustments will occur over a 30-year hydrogen market expansion period, the average yearly quantity adjustment for a commodity will be less than 0.1 of one percent. Similar considerations apply to all countries adopting hydrogen. On-going globalization effects on trade over the coming years seem bound to overwhelm the small trade adjustments to the reduction in oil imports.

A similar conclusion applies to effects on world capital markets. With falling OPEC incomes, OPEC countries' purchases of U.S. assets would be reduced. Overall financial lending to industrialized countries would be lowered. Any effect on world interest rates is unlikely to be perceptible given that OPEC demand for financial assets is a small part of total world demand for assets.

Recommendations

This study identified possible employment impacts that could result from hydrogen market expansion in the transportation and stationary and portable power sectors. As noted previously, any study of potential future impacts necessarily presents difficult challenges and involves significant uncertainties. Results and recommendations should be considered with those issues in mind. The scenarios, hydrogen generation options, and regions selected for the study yielded a reasonable measure of the potential opportunities that hydrogen presents to U.S. employment. The study considers introduction of the fuel cell vehicle and supporting hydrogen infrastructure development that spans a period of approximately 40-50 years. The study highlights possible skill and education needs to support the associated industries and technologies. In addition to the specific skill requirements of the fuel cell industry, future education of the next generation should be focused on skill sets that have the ability to adapt to changing technologies.

Training implications of the job changes vary by industry and skill, as indicated by a survey of industry opinions. Most of the needs for new skills can be supplied by normal rates of new entry into the labor force as workers receive training in new, hydrogen-related skills. Considering the small proportion of all engineers in the U.S. that the automobile industry employs, engineering schools have the capacity to respond to changes in the fields of engineering.

Following are employment-related recommendations for a transition to a hydrogen economy.

1. Training programs

a. Training and retraining programs may be needed to help ensure that the U.S. workforce possesses appropriate skills and that sufficient numbers of trained personnel are available to meet the manufacturing requirements at the time that hydrogen fuel cell vehicles begin to come off the assembly lines. Development of these programs should involve close coordination between the fuel providers and auto manufacturers, and schools. At the appropriate time, university and vocational programs will need to be assessed to understand where opportunities lie and what additional curricula may be needed.
b. Training and retraining programs may be useful in related aftermarket areas such as repair and recycling.
c. Educational programs aimed at the general public could help to influence people to pursue jobs in the hydrogen and fuel cells industries.

2. Additional analysis

a. *Analysis of training needs*: The study found that training for new skills may be needed across a wide spectrum of industries. Most of the needs for new skills can be supplied by normal rates of new entry into the labor force as workers receive training in new, hydrogen-related skills. Some changes in skills appear to be relatively well defined, but many likely changes remain difficult to forecast, since many of the technologies are still maturing. Many job tasks remain unknown at present, making identification of training needs an interactive task with job definition.
b. *Assessment of skill changes, with attention to industry adjustments in different regions*: Unemployment resulting from the elimination of jobs associated with obsolete technologies and industries is estimated to be slight. The supply of labor with technology-specific skill sets may or may not keep pace with the labor demands associated with the rapid growth in hydrogen-related industries. Even though this study indicated every region would gain jobs, surpluses and shortages of skilled workers could vary regionally.

End Notes

[1] The study was reviewed by and benefited from the advice and comments of the members of an industry advisory panel including Dr. John Johnston, former Planning Executive, Corporate Strategic Research Lab, Exxon-Mobil Research and Engineering; Dr. Alan Lloyd, President of the International Council on Clean Transportation; Dr. Walter McManus, Director Automotive Analysis Division, University of Michigan Transportation Research Institute; Mr. Gregory Morris, Senior Vice President, HydroGen, LLC and Executive Director, Cullen Engineering Research Foundation; and Dr. Robert Rose, Executive Director, U.S. Fuel Cell Council.

[2] DOE initiated its Hydrogen Analysis (H2A) initiative in February 2003 as a means of establishing a standard format and list parameters for estimating and comparing the lifecycle costs of hydrogen production and delivery technologies. See, http://www.hydrogen.energy.

[3] President Bush announced the HFI in his 2003 State of the Union Address. The HFI consists of a $1.2 billion program to develop commercially viable hydrogen fuel cell and infrastructure technologies by the year 2020.

[4] Government Performance and Results Act, Public Law No. 103-62. This law requires Federal agencies to develop and report annually on performance measures and goals for each program activity of the agency. The 2007 report was prepared by DOE and the National Renewable Energy Laboratory and is entitled, Projected Benefits of Federal Efficiency and Renewable Energy Programs, FY 2007 Budget Request (March 2006) See, http://www1.eere.energy.gov/ba/pba/2007_benefits.html. (Projected Benefits FY 2007)

[5] For example, the IMPLAN model has been used for long term technology assessments in the following reports: Robert H. Beach and Martin T. Ross, *General Equilibrium Assessment of Regional Climate Change Policy*, Proceedings of the 2004 National IMPLAN User's Conference (MIG, Inc., Stillwater, MN (Oct. 2004) at 151-168; and Northwest Economic Associates, *Assessing the Economic Development Impacts of Wind Power* (Feb. 12, 2003).

Chapter 1

INTRODUCTION

This Department of Energy (DOE or Department) study is in response to Section 1820 of the Energy Policy Act of 2005 (Public Law 109-58) (EPACT). Section 1820, "Overall Employment in a Hydrogen Economy," requires the Secretary of Energy (Secretary) to carry out a study of the effects of a transition to a hydrogen economy on overall employment in the United States.[1] As required by Section 1820, the present report considers:

Replacement effects of new goods and services
International competition
Workforce training requirements
Multiple possible fuel cycles, including usage of raw materials
Rates of market penetration of technologies
Regional variations based on geography
Specific recommendations of the study

Both the Administration's National Energy Policy[2] and the Department's Strategic Plan[3] call for reducing U.S. reliance on imported oil and reducing greenhouse gas emissions. The National Energy Policy also acknowledges the need to increase energy supplies and use more energy-efficient technologies and practices. President Bush proposed in his January 2003 State of the Union Address to advance research on hydrogen so that it has the potential to play a major role in America's future energy system. Consistent with these aims, EPACT authorizes a research, development, and demonstration program for hydrogen and fuel cell technology.

Section 2 of this report considers a base case and two hydrogen adoption scenarios. The base case assumes no broad-based use of hydrogen across the economy. The first hydrogen adoption scenario uses a market penetration rate consistent with the Hydrogen Fuel Initiative (HFI), introduced by President Bush in his 2003 State of the Union Address (the "HFI Scenario"). The HFI consists of programs that focus on the development of commercially viable hydrogen fuel cell and infrastructure technologies by the year 2020. In the HFI Scenario, market penetration begins in 2018, with full market penetration achieved by 2050. The second hydrogen adoption scenario also begins in 2018 but, applying the DOE's analysis supporting its 2007 program benefits estimation[4] assumes a less aggressive rate of market penetration by hydrogen technologies (the "Less Aggressive Scenario"). In estimating the effects of the base case and the two hydrogen adoption scenarios on overall employment and the economy, the study uses a modeling tool called IMPLAN. IMPLAN, an acronym for "IMpact Analysis for PLANning," is a methodology that has been used in other contexts to predict the economic impacts of long-term technological developments.[5] Specifically, the study utilizes a 509-sector input-output IMPLAN model of the U.S. economy based on the most recent available U.S. input-output table (2002). Furthermore the mix of hydrogen production technologies considered in the study spans a range of the most likely production scales and feedstock options. Most of the technologies used in the study are relatively mature, have been intensively studied, or both. Thus, the cost studies that have been completed are based on conservative assumptions regarding the technologies that will be available under the various scenarios. The studies accordingly provide equally conservative estimates of future costs and the detail needed for input to the IMPLAN model. The IMPLAN methodology is described in greater detail in Section 3 and Appendix 4 of the study.

In order to identify the impacts of a hydrogen economy, one must look far enough into the future for that economy to be well underway. For this study, the year 2050 was selected as a target because DOE's own analysis using its Hydrogen Analysis (H2A) model[6] for comparing the lifecycle costs of different hydrogen technologies indicated that development of the hydrogen economy could be well underway and the impacts readily apparent by that year. Key parameters were extrapolated from 2030 (the latest year of the Annual Energy Outlook forecasts) to 2050. Two other years were selected for detailed analysis – 2020, when new hydrogen industry sectors begin to appear in the two hydrogen scenarios, and 2035, when those sectors are likely to reach the critical mass needed to identify them as belonging to a nascent hydrogen economy.

Section 3 gives estimates of employment effects of the transition to hydrogen technologies at the national level in the years 2020, 2035 and 2050. Using a model of the U.S. economy, estimates are presented of: (1) the net effect on total employment during and after market penetration, (2) employment creation and replacement of jobs and (3) education, training and re-training needs.

Section 4 considers impacts in five geographic regions of the country: the Upper Midwest, the combined area of Lower New England and the Upper Mid-Atlantic, California, Tennessee and metropolitan Houston.

Section 5 considers the international position of the U.S. as it relates to hydrogen. Estimates are given of impacts on the U.S. share of world production of vehicles, vehicle parts and other products using hydrogen, and on oil and natural gas imports. The role of multinational corporations in hydrogen markets is assessed. The nature and importance of repercussions through the balance of payments and the dollar exchange rate are also analyzed.

Section 6 reprises the conclusions and recommendations resulting from the study.

This report is being disseminated by the Department of Energy. As such, the document was prepared in compliance with Section 515 of the Treasury and General Government Appropriations Act for Fiscal Year 2001 (Public Law 106-554) and information quality guidelines issued by the Department of Energy. The study was reviewed both internally and externally prior to publication. For purposes of external review, the study benefited from the advice and comments of the members of an independent advisory panel. That panel included Dr. John Johnston, former Planning Executive, Corporate Strategic Research Lab, Exxon-Mobil Research and Engineering; Dr. Alan Lloyd, President of the International Council on Clean Transportation; Dr. Walter McManus, Director, Automotive Analysis Division, University of Michigan Transportation Research Institute; Mr. Gregory Morris, Senior Vice President, HydroGen, LLC and Executive Director, U.S. Fuel Cell Council.

End Notes

[1] U.S. Congress. *The Energy Policy Act of 2005*. Public Law No. 109-58 (Aug. 8, 2005) (Washington, D.C.: USGPO, 2005), Sec. 1820(a), 119 Stat. 594, 1132.

[2] *National Energy Policy; Report of the National Energy Policy Development Group* (Washington, D.C.: USGPO, May 2001), pp. 2-9, 3-11.

[3] U.S. Department of Energy, Office of Program Analysis and Evaluation, *Department of Energy Strategic Plan*. 2006, p. 9.

[4] Projected Benefits FY 2007, *supra* note 4.

[5] Minnesota IMPLAN Group, Inc., *IMPLAN Professional, Version 2.0, Social Accounting & Impact Analysis Software, User Guide, Analysis Guide*, and Data Guide, 3rd Edition Stillwater, Minn., February 2004.

[6] DOE initiated its H2A initiative in February 2003 as a means of establishing a standard format and list of parameters for estimating and comparing the lifecycle costs of hydrogen production and delivery technologies. See, http://www.hydrogen.energy.gov/h2a_analysis.html.

Chapter 2

SCENARIOS SHAPING A HYDROGEN ECONOMY

As indicated above, three scenarios were defined to provide a context for the analysis and an internally consistent set of assumptions for analyzing the employment impacts of a hydrogen economy. These include:

a base or reference case,

the HFI Scenario, an aggressive hydrogen market development case modeled on President Bush's **Hydrogen Fuel** Initiative (HFI) as outlined in his 2003 State-of-the-Union address, referred to in this report as the HFI, and

the Less Aggressive Scenario, an intermediate market development case modeled on supporting analyses from the DOE 2006 report satisfying requirements of the program benefits estimation,[1]

Each of the scenarios is summarized below.

2.1. BASE CASE

The base case represents a world in which hydrogen has only a minimal role as a primary energy carrier, thereby providing a point of reference against which to measure the impact of the hydrogen economy.While small fleets of hydrogen vehicles are operated as part of technology validation and other R&D projects and small quantities of hydrogen fuel are supplied, primarily from existing industrial gas production facilities, the base case assumes that there is no widespread demand for hydrogen fuel or vehicles.For purposes of analysis, the high oil price case in the Energy Information Administration's **2006 Annual Energy Outlook** (AEO-06) was selected to represent such the future under the base case.[2]This base, or high oil price, case reflects a continuation of recent demographic, economic and oil price trends.

Table 2.1. Key Economic and Demographic Features of the Base Case

Parameter		2005	2020	2035	2050	Average % change/yr (2005-2050)
Population (millions)		296.8	337.0	378.1	420.1	0.78%
Employment (millions)		131.6	147.3	164.8	184.4	0.75%
Driving age population (millions)		231.7	265.3	298.5	331.4	0.80%
GDP (billions of 2005 $)		12,566	19,693	29,249	38,342	2.48%
Gasoline (2005 $/gal)		2.38	3.01	3.38	3.56	0.90%
Crude oil (2005 $/bbl)a		55.93	85.06	104.50	117.27	1.65%
Light-duty vehicles (millions)		223.4	286.6	339.1	386.6	1.22%
Light-duty-vehicle fuel use	quads/yr[b]	16.0	19.9	23.6	25.6	1.04%
	billions of gallons of gasoline equivalent/yr	135.6	168.6	199.9	216.2	1.04%

[a] Imported low-sulfur light crude oil; [b] 1 quad = 1 quadrillion Btu.

Sources: *Population, GDP, crude oil price, light-duty vehicles, travel and fuel use 2005-2030* : *Annual Energy Outlook 2006, with Projections to 2030*, U.S. Department of Energy, Energy Information Administration report DOE/EIA-0383(2006), February 2006, accessed August-September 2006 at http://www.eia.doe.gov/oiaf/aeo/aeohighprice.html.

Population, GDP, crude oil price 2030-2050: extrapolated from AEO-06 high price forecast

Employment: Quarterly Census of Employment and Wages, U.S. Department of Commerce, Bureau of Labor Statistics, accessed August-September 2006 at http://www.bls.doc.gov/cew/home.htm.

Gasoline price: Adjusted from the AEO-06 high price forecast to reflect long-term relationships between crude oil and delivered gasoline prices and extrapolated to 2050.

Table 2.1 shows the values and associated growth rates of selected economic and demographic parameters under the base case.Beyond 2030, energy prices and economic and demographic variables are based on trends from the Annual Energy Outlook, while fuel use is estimated using the VISION model (described in Appendix 1).[3]Economic output rises at approximately 2.5% per year from 2005 to 2050. The population grows much less rapidly (at approximately 0.8% per year), while the number of light-duty vehicles and vehicular travel rise at 1.2% and 1.4% per year, respectively.Because of steady increases in the world oil price (which reaches over $117 per barrel in 2050), conventional light-duty vehicles become more fuel-efficient and fuel use grows somewhat more slowly, at 1% per year.

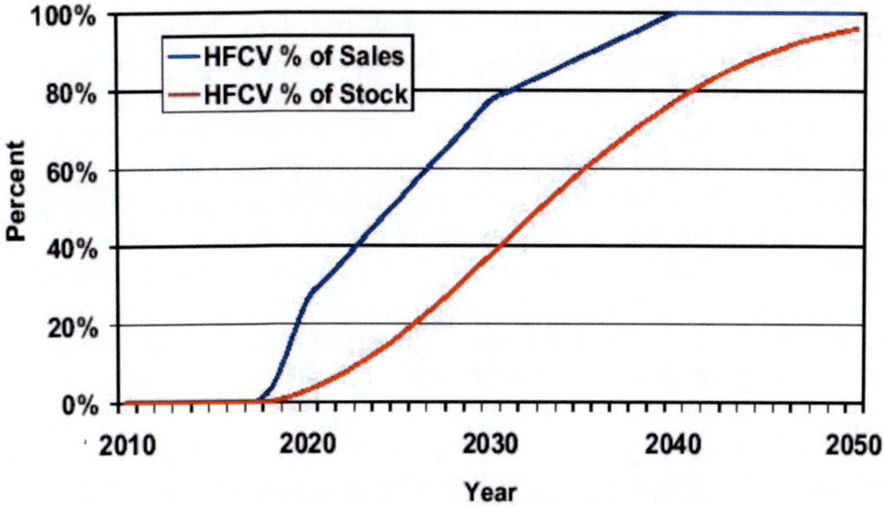

Sources *Hydrogen Posture Plan*, U.S. Department of Energy report, February 2004, August-September 2006 at http://www1.eere.energy.gov (for sales data); Stock data were simulated by the VISION model, calibrated to reflect parameters of AEO 2006 high price case.

Figure 2.1. Hydrogen Fuel Cell Vehicles (HFCVs) as a Percent of Light-Duty Vehicles Sold and Total Light-Duty Vehicle Stock in the **President's Initiative**

2.2. Hydrogen Use in the HFI Scenario

Compared with the base case in which hydrogen vehicles achieve no more than token market penetration, the HFI presents a vision of rapid market development.[4]Following achievement of the program's **technical targets** for hydrogen, fuel cells and infrastructure technologies, hydrogen fuel cell vehicles are introduced to the general public around 2018.Hydrogen vehicle production increases rapidly, capturing all segments of the light-duty-vehicle market. As older, conventional vehicles are retired from use, nearly all light-duty vehicles are hydrogen-fueled by 2050, as shown in Figure 2.1.

Under the HFI Scenario, fuel infrastructure is installed in lock-step with the growing population of hydrogen-fueled vehicles, thereby providing both the supply chain for a growing hydrogen fuel industry and a push to propel ever-higher hydrogen vehicle sales.Initially, most hydrogen is produced on-site, in relatively small, "distributed" facilities or delivered to fueling stations from existing, large industrial plants.Though similar to gasoline stations in outward appearance, distributed facilities are more costly to build and maintain.Thus, as demand for hydrogen rises, fuel production shifts to centralized facilities that can better capture economies of scale.A mix of centralized hydrogen production technologies arises as different regions capitalize on local resources.Regions with nearby coal supplies and CO_2 sequestration sites opt for more coal gasification while those with ample wind or biomass favor hydrogen production technologies that rely more on those resources.From centralized production facilities, hydrogen is delivered to local fuel stations where it is dispensed along with conventional motor fuels.As demand increases, delivery technologies shift from primarily tanker trucks carrying cryogenic liquid hydrogen to a mix of trucks carrying cold compressed hydrogen in insulated high-pressure tanks, and, beginning between 2020 and 2035, gas pipelines. By 2050, when hydrogen has replaced nearly all motor gasoline, most local fuel stations no longer dispense gasoline.

The development of an on-board hydrogen storage technology for hydrogen-powered vehicles has implications for the hydrogen delivery infrastructure. Currently, compressed hydrogen storage is the dominant approach for on-board hydrogen storage, with most systems using 5,000 psi composite tanks. Next generation systems, which have started to appear in demonstrations use higher pressure 10,000 psi tanks.

Consequently, under the HFI Scenario, stations will need high-pressure compressors and delivery systems.The HFI Scenario also reflects the fact that, even today, two other general storage approaches are being investigated: solid-state and chemical hydrogen storage systems. Solid-state on-board storage encompasses complex metal hydrides and carbon-based materials. The infrastructure for solid-state storage will likely have somewhat lower pressurization requirements, but will probably require additional thermal management systems to cool the solid-state material and on-board storage system (and potentially the hydrogen fuel itself) during the fill process. A third approach to on-board storage includes chemical hydrogen (e.g., chemical hydrides, hydrocarbon compounds) that will have to be regenerated off-board. Off-board regeneration has significant infrastructure implications involving return of the spent material to central facilities and the energy needed for regeneration. The current study assumes compressed hydrogen storage on the vehicle only.

For the HFI Scenario, it is assumed that by 2050, 55% of all hydrogen is produced from coal with carbon sequestration, 20% from biomass, 15% from wind, and 10% from nuclear processes.These are estimated national averages. The mix for individual regions (and indeed for individual states and localities within regions) is likely to vary considerably.

2.3. Hydrogen Use in the Less Aggressive Scenario

The Less Aggressive Scenario represents a future midway between the optimism of the HFI Scenario and the status quo of the base case. [5]Like the HFI, the Less Aggressive Scenario superimposes a set of hydrogen fuel cell market penetration assumptions on the economic and demographic features of the base case.However, unlike the HFI Scenario, hydrogen vehicles come on the market somewhat later in the Less Aggressive Scenario and, amid stiff competition from hybrids, plug-ins, flex-fuel vehicles, and advanced conventional vehicles, gain market share at a slower pace, as shown in Figure 2.2.

Though hydrogen technologies are slower to gain acceptance under the Less Aggressive Scenario than the HFI Scenario,demand nevertheless grows quickly under the Less Aggressive Scenario relative to historic standards.Thus, the same shift from distributed to centralized hydrogen production and the same mix of delivery technologies in 2020, 2035 and 2050 as in the HFI Scenario are assumed.

2.4. Hydrogen Vehicles

Table 2.2 summarizes the fuel cell and hydrogen storage cost targets of the DOE Hydrogen Program.[6]Both the HFI and Less Aggressive Scenarios assume achievement of all targets.The resulting fuel cell vehicle costs roughly the same as the gasoline-fueled vehicle.Achievement of the targets also results in a fuel cost comparable to a gasoline vehicle on a $/mile-traveled basis.

In all scenarios, sales and stock of light-duty vehicles grow, and the stock of all light-duty vehicles grows faster than population grows. As a result, the number of vehicles per person and the number per driving-age person continue to rise (from 0.75 to 0.92 vehicles per capita and from 0.96 to 1.17 vehicles per driving-age population).Saturation (i.e., ratios approaching some natural limit) does not occur in these ratios, as indeed it has not occurred with respect to vehicles per household or vehicles per licensed driver. With expanding sales of total light-duty vehicles, the market penetration rates of the two hydrogen scenarios result in significant numbers of hydrogen vehicles on the road by 2030.

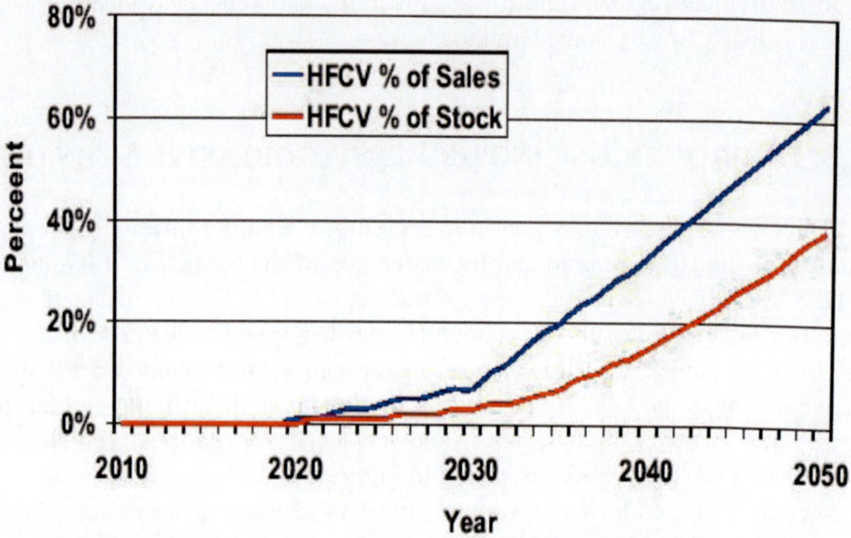

Sources: sales: *FY 2007 Benefits Estimates*, U.S. Department of Energy accessed August-September 2006 at http://www1.eere.energy.gov/ba/pba/2007_benefits.html; Stock: calculated by VISION model, calibrated to reflect parameters of AEO 2006 high price case.

Figure 2.2. Hydrogen Fuel Cell Vehicles (HFCVs) as a Percent of Light-Duty Vehicles Sold and Total Light-Duty Vehicle Stock in the Less Aggressive Scenario

Table 2.2. Selected Fuel Cell Vehicle Cost Targets (2005 Prices)

Component	Unit	2004 Status	2015 Target
Fuel cell stack[a]	$/kW_e$	75	20
Fuel cell system[a]	$/kW_e$	120	30
On-board storage system	$/kWh net	>6	2

[a] Based on high volume (500,000 stacks per year).

Source: U.S. Department of Energy, *Hydrogen, Fuel Cells & Infrastructure Technologies Program, Multi-Year Research, Development and Demonstration Plan,* February 2005, accessed September 2006 at http://www1.eere.energy.

2.4.1. Hydrogen Vehicles in the HFI Scenario

In this report, the market penetration assumptions of the HFI Scenario (Figure 2.1) were selected as the upper bound or most aggressive market penetration scenario.As shown in Table 2.3, hydrogen fuel cell vehicles account for 27% of new sales and 3% of light-duty vehicles on the road in 2020, 89% of sales and 60% of vehicles on the road in 2035, and all light-duty-vehicle sales and 96% of the light-duty vehicle stock by 2050.This translates into nearly 24 million vehicles sold and nearly 350 million hydrogen vehicles on the road in 2050.

Figure 2.3a and 2.3b display the distribution of fuel use by light-duty vehicles under the HFI Scenario. As the stock of conventional vehicles is displaced by fuel cell vehicles, gasoline consumption declines.Total fuel use also declines as a result of the increased fuel efficiency of hydrogen fuel cell vehicles relative to conventional vehicles.Estimates of the production cost of light-duty hydrogen fuel cell vehicles are reported in Appendix 3.

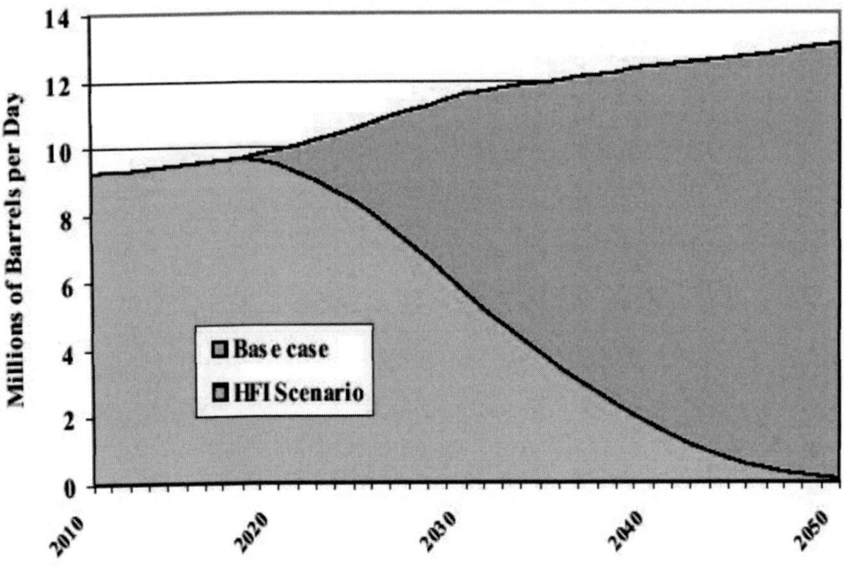

Source: calculated by VISION model using AEO 2006 high price calibration and market penetration assumptions of the HFI.

Figure 2.3a. Oil Consumption, Base Case vs. HFI Scenario (in millions of barrels of oil per day)

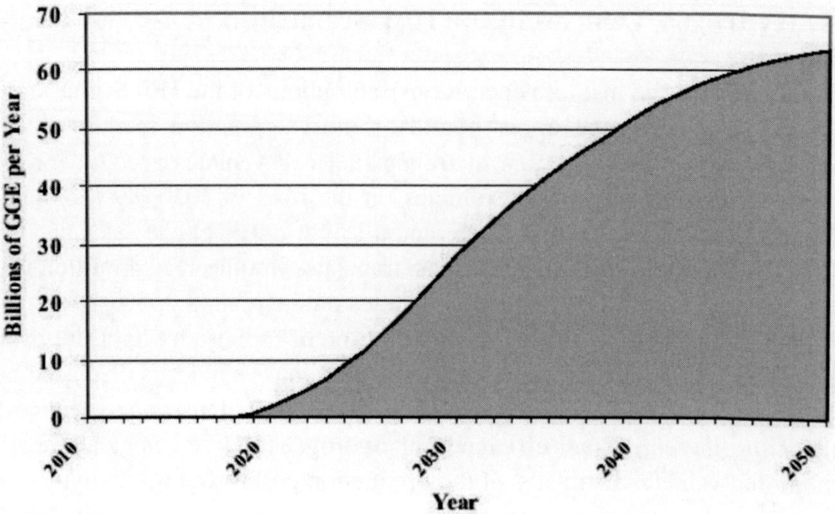

Source: calculated by VISION model using AEO 2006 high price calibration and market penetration assumptions of the HFI.

Figure 2.3b. Hydrogen Consumption, HFI Scenario (in billions of gallons of gasoline equivalent per year)

Table 2.3. Hydrogen Fuel Cell Vehicles in the HFI Scenario

Year		**2020**	**2035**	**2050**
Sales	millions of light-duty vehicles sold/yr	4.9	19.5	23.9
Stock	millions of light-duty vehicles in use	8.8	197.5	347.5
	% of all light-duty vehicles sold	26.6%	88.9%	100%
Hydrogen fuel use	% of all light-duty vehicles in use	3.1%	59.6%	96.0%
	quads/yr[a]	0.22	5.24	8.02
	billion gge/yr[b]	1.8	41.9	64.1

[a]1 quad = 1 quadrillion Btu; [b]gge = gallons of gasoline equivalent.

Source: Calculated by VISION model using AEO 2006 high price calibration and market penetration assumptions of the HFI.

2.4.2. Hydrogen Vehicles in the Less Aggressive Scenario

Figure 2.2 illustrates the hydrogen fuel-cell-vehicle market penetration assumptions and the resulting shares of total light-duty vehicles in the Less Aggressive Scenario.Compared with the HFI Scenario (in which hydrogen vehicles gain complete market dominance by 2050), hydrogen vehicles capture 63% of sales and represent 38% of light-duty vehicles in use in the Less Aggressive Scenario.As shown in Table 2.4, these shares represent the sale of 15 million hydrogen vehicles and 144 million hydrogen vehicles on the road in 2050.The 144 million hydrogen vehicles on the road are expected to consume approximately 3.7 quads (3.7 quadrillion Btu) of hydrogen fuel per year.Given the vehicle and fuel economy assumptions of the Less Aggressive Scenario, hydrogen vehicles displace 11.9 quads of oil (nearly 6.6 million bbls/day) in 2050 under the Less Aggressive Scenario.

As shown in Figure 2.4a and 2.4b, hydrogen's share of light-duty vehicle energy use is substantially lower in the Less Aggressive Scenario than in the HFI Scenario.Not only is there less hydrogen fuel use (due to fewer fuel cell vehicles in operation by 2050), but total energy use is substantially higher under the Less Aggressive Scenario.The latter is due to a combination of more conventional vehicles on the road and the relatively lower fuel economy of those vehicles, compared with hydrogen vehicles.

Table 2.4. Hydrogen Fuel Cell Vehicles in the Less Aggressive Scenario

Year		**2020**	**2035**	**2050**
Sales	millions of light-duty vehicles sold/yr	0.2	4.4	15.1
	% of all light-duty vehicles sold	0.8%	20.1%	63%
Stock	millions of light-duty vehicles in use	1.4	25.3	144
	% of all light-duty vehicles in use	0.5%	7.5%	38.2%
Hydrogen fuel use	quads/yr	0.04	0.70	3.73
	billion gge/yra	0.32	5.59	29.95

[a]gge = gallons of gasoline equivalent.

Source: Calculated by VISION model using AEO 2006 high price calibration and market penetration assumptions of the Less Aggressive scenario.

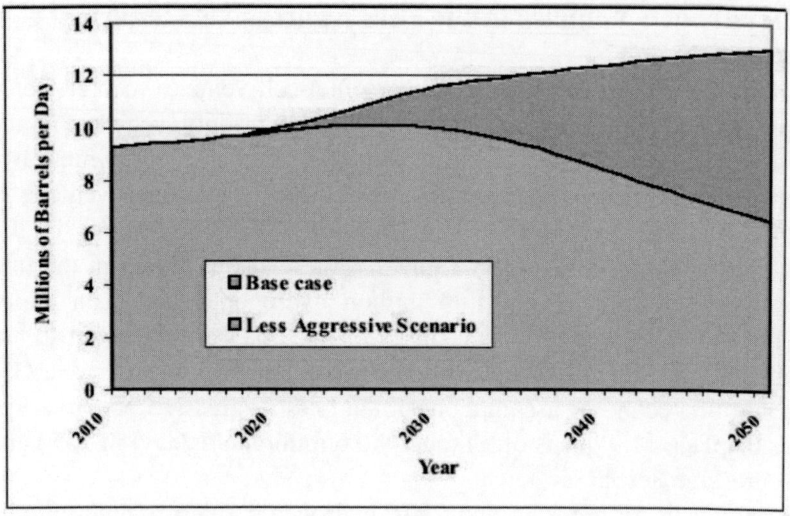

Source: calculated by VISION model using AEO 2006 high price calibration and market penetration assumptions of the Less Aggressive scenario.

Figure 2.4a. Oil Consumption, Base Case vs. Less Aggressive Scenario (in millions of barrels of oil per day)

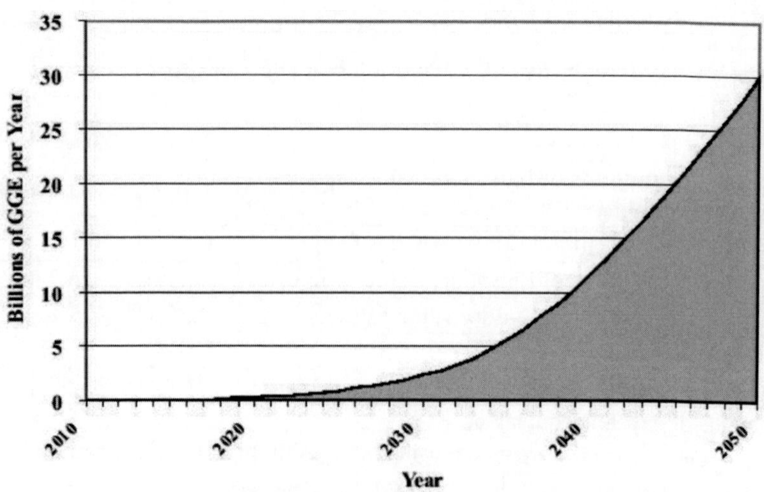

Source: calculated by VISION model using AEO 2006 high price calibration and market penetration assumptions of the Less Aggressive scenario.

Figure 2.4b. Hydrogen Consumption, Less Aggressive Scenario (in billions of gallons of gasoline equivalent per year)

2.5. Other Transportation Markets

This report focuses on light-duty vehicles since fuel substitution in this market will have the greatest impact on reducing oil consumption in the transportation sector. However, hydrogen fuel cell technologies will likely have applications beyond the light-duty fleet. For example, transit and school buses are important potential markets. Transit buses are a relatively small market with high utilization and turnover. School buses are a larger market, but with relatively low utilization and slow turnover. Together, these markets could add 1% to the total value of light-duty vehicle sales in 2050.

2.6. Stationary Fuel Cells

In both hydrogen adoption scenarios, stationary fuel cells initially provide power, including back-up power, for remote locations not easily served by the electric power transmission and distribution grid. Following success in these markets, fuel cells begin to penetrate markets for portable power, then markets for all types of distributed power.For back-up and remote power markets, users are willing to pay a premium for secure, reliable electricity.Such markets include hospitals, hotels, data centers, and computer facilities, where uninterrupted power is critical. While these initial markets provide the sales volumes to launch the industry, the fuel cells themselves operate relatively few hours per year and displace little grid electricity.

Portable power is a likely follow-on market for fuel cells.Fuel cells are being eyed for a wide array of portable applications ranging from consumer electronics to small-scale power production.Portable electronics (e.g., cell phones and notebook computers) utilizing premium lithium-ion and nickel metal-hydride batteries are particularly promising candidates for substitution by fuel cells.Premium batteries are popular because of their high energy density.As portable devices become more complex and their power requirements increase, fuel cells become an increasingly promising alternative.

Since 2002, the number of devices containing premium batteries that the average American carries has grown 10% annually, to 0.5 devices per person.[7]As the variety and functionality of portable consumer electronics grow, each person will carry increasingly multi-functional devices.While this will increase the demand for portable power, it may not increase the total number of devices beyond some natural limit of perhaps 1 device per person on average.Assuming the base-case population forecast of 420 million, some 400 million portable fuel cells could be in use in 2050.

As fuel cells penetrate the portable power market and enter the broad residential and commercial power market, electricity displacement becomes substantial. Like fuel cells for mobile applications, stationary fuel cells are assumed to meet program cost and performance targets. In this study, stationary fuel cells are assumed to be natural gas fueled, where the fuel cell system includes an integrated reformer to produce the hydrogen.

2.6.1. Stationary Fuel Cells in the HFI Scenario

In the HFI Scenario, stationary fuel cells are assumed to achieve 1% penetration of "new" electric demand in 2020, 5% in 2035, and 10% in 2050.[8] "New" demand is defined as the difference between electricity demand in 2015 and the analysis year (2020, 2035 or 2050), and is meant to include a variety of markets – back-up power, portable power and the broader market for residential and commercial power – which increasingly shift to fuel cells. Penetration by fuel cells results in displacement of 0.01 quad of grid electricity in 2020, 0.3 quads in 2035, and slightly over 1 quad in 2050.

2.6.2. Stationary Fuel Cells in the Less Aggressive Scenario

As in the HFI Scenario, stationary fuel cells are assumed to first penetrate remote and back-up power markets, followed by portable power and the broad market for residential and commercial electricity.Large-scale deployment is assumed to begin in 2020 and to proceed at roughly half the rate that applied in the HFI Scenario. Thus, stationary fuel cells achieve 2% penetration of new electric demand in 2035 and 5% in 2050, thereby displacing approximately 0.1 quad of electricity in 2035 and 0.5 quads in 2050.As in the HFI Scenario, program cost and performance targets are assumed to be met.

2.7. Hydrogen Production and Delivery in the Two Hydrogen Adoption Scenarios

In addition to employment generated (or displaced) by hydrogen vehicles replacing conventional vehicles and by fuel cells replacing stationary power, the infrastructure to produce and distribute hydrogen will be an important source of employment impacts. To model these impacts *vis-à-vis* the base case, candidate production and delivery technologies were identified and characterized in sufficient detail to provide estimates of (1) the size and number of facilities required to supply a given quantity of hydrogen fuel, (2) when those facilities might become widely available, (3) required feedstock and other inputs (including necessary expansion of feedstock transportation and other supply chain infrastructure), and (4) capital and operating costs.These costs were then used to characterize new industries in the IMPLAN model.

The following production technologies were included in the analysis:

Distributed reforming of natural gas
Centralized reforming of natural gas, with carbon capture and sequestration[9]
Distributed electrolysis
Coal gasification, with carbon capture and sequestration
Centralized biomass gasification
Nuclear electrolysis
Nuclear thermo-chemical water splitting
Centralized wind electrolysis

Although most of these technologies could be of several designs, a single representative design was characterized for each.The H2A model, a tool developed by DOE with industry assistance, was used for this purpose.[10]Further detail on that tool and key parameters assumed in the analysis are contained in Appendix 1.

As stated above, the same shift from distributed to centralized production and the same mix of hydrogen production and delivery technologies are assumed for the HFI and the Less Aggressive Scenarios. Tables 2.5-2.7 list the production technologies included in these scenarios, along with the number of plants required for each scenario for the three analysis years.Estimates of the capital, construction, and operating costs of these facilities are reported in Appendix 2.

Table 2.5. Unit Size, Number of Units and Percent Hydrogen Productionby Technology and Scenario – 2020

Hydrogen Production Technology	Unit size (T/d)[a]	% of Hydrogen Produced[b]	Units in HFI[b]	Units in Less Aggressive Scenario[b]
Distributed reforming of natural gas	1.5	30%	1,319	235
Distributed electrolysis	1.5	10%	440	79
Wind electrolysis	86	5%	3	1
Nuclear electrolysis	719	0%	0	0
Central reforming of natural gas	380	20%	3	1
Biomass gasification	164	10%	4	1
Coal gasification	246	25%	5	1
Nuclear thermochemical	768	0	0	0

[a] Metric tons/day.

[b] Production shares and number of units are estimated from hydrogen demand under each scenario and facility construction schedules.

Table 2.6. Unit Size, Number of Units and Percent Hydrogen Production by Technology and Scenario – 2035

Hydrogen Production Technology	Unit size (T/d)[a]	% of Hydrogen Produced[b]	Units in HFI[b]	Units in Less Aggressive Scenario[b]
Distributed reforming of natural gas	1.5	2%	2,073	277
Distributed electrolysis	1.5	1%	1,036	139
Wind electrolysis	86	15%	156	22
Nuclear electrolysis	719	7%	14	2
Central reforming of natural gas	380	1%	4	1
Biomass gasification	164	19%	156	21
Coal gasification	246	50%	224	30
Nuclear thermochemical	768	5%	9	2

[a] Metric tons/day.

[b] Production shares and number of units are estimated from hydrogen demand under each scenario and facility construction schedules.

Similarly, delivery technologies were characterized at a level of detail sufficient to produce estimates of total fixed and variable costs under the two hydrogen scenarios.As with production technologies, delivery technologies were characterized using a DOE-developed H2A tool.[11]

Table 2.8 lists the delivery technologies included in this analysis and the shares of total production served by each.Totals add to the share of hydrogen produced in central plants, not 100%; production at local refueling stations ("distributed production") accounts for the remainder.

Table 2.7. Unit Size, Number of Units and Percent Hydrogen Productionby Technology and Scenario – 2050

Hydrogen Production Technology	Unit size (T/d)[a]	% of Hydrogen Produced[b]	Units in HFI[b]	Units in Less Aggressive Scenario[b]
Distributed reforming of natural gas	1.5	1%	1,586	741
Distributed electrolysis	1.5	0.5%	794	370
Wind electrolysis	86	15%	243	114
Nuclear electrolysis	719	4.5%	13	7
Central reforming of natural gas	380	0.5%	4	2
Biomass gasification	164	20%	251	118
Coal gasification	246	53%	363	170
Nuclear thermochemical	768	5.5%	14	7

[a] Metric tons/day.

[b] Production shares and number of units are estimated from hydrogen demand under each scenario and facility construction schedules.

Table 2.8. Shares[a] of Total Hydrogen Production, by Delivery Mode, for the HFI and the Less Aggressive Scenarios, by Year

Share by Delivery Mode (%)	2020	2035	2050
Pipeline	0	55	70
Compressed gas truck	25	24	17
Cryogenic liquid truck	35	18	11.5
Total delivered	55	97	98.5
Distributed production	43	3	1.5

[a] Estimated as a function of hydrogen demand and infrastructure construction schedules.

As with hydrogen production, hydrogen distribution can be accomplished by various technologies. For this analysis, distribution (i.e., delivery) is assumed to occur via liquid or compressed gas truck or gas pipeline. These alternatives represent relatively mature technologies for which cost models have been developed and a considerable body of analysis has been completed. The delivery cost models were developed as part of the H2A project to be consistent in form and approach to the production cost models mentioned above. Both sets of H2A models (delivery and production) are discussed in Appendix 2. Research and development are continuing on other delivery technologies that ultimately may be less costly to implement and more efficient to operate and maintain than those assumed in this report. These technologies include hydrogen carriers, adsorbing materials and other novel processes, any or all of which could play an important role in hydrogen delivery.As with the choice of production technologies, determining the most realistic mix of future delivery options is a study in and of itself and will depend heavily on research and development accomplishments in the coming years.

End Notes

[1] Projected Benefits FY 2007, *supra* note 4.

[2] *Annual Energy Outlook 2006, with Projections to 2030*, U.S. Department of Energy, Energy Information Administration report DOE/EIA-0383(2006), Feb. 2006, accessed Aug.-Sept. 2006 at http://www.eia.doe.gov/oiaf/aeo/aeohighprice.html.

[3] *VISION 2006 AEO High Price Base Case*, personal communications from A. Vyas, Argonne National Laboratory, recalibrated to AEO 2006 high price case from posted AEO 2006 base case, accessed August-September 2006 at http://www.transportation.anl.gov/software/VISION/index.html.

[4] *Hydrogen Posture Plan*, U.S. Department of Energy report, February 2004, accessed August-September 2006 at http://www1.eere.energy.gov/hydrogenandfuelcells/posture_plan04.html.

[5]Projected Benefits FY2007, supra note 4.

[6] U.S. Department of Energy, *Hydrogen, Fuel Cells & Infrastructure Technologies Program, Multi-Year Research, Development and Demonstration Plan*, February 2005, accessed September 2006 at http://www1.eere.energy.gov/hydrogenandfuelcells/mypp.

[7] Darnell Group, *Fuel Cells for Portable Power:Markets, Manufacturer and Cost,* January 13, 2003.

[8] U.S. DOE, Energy Information Administration, Annual Energy Outlook 2005, Table 8.8.

[9] Although this technology is available for hydrogen production, it is not being pursued in DOE's R&D program.

[10] *H2A Production Analysis,* U.S. Department of Energy, Central and distributed production analysis tools accessed August-September 2006 at http://www.hydrogen.

[11] *H2A Delivery Analysis,* U.S. Department of Energy, accessed August – September 2006 at http:// www.hydrogen.energy.gov/h2a_delivery.html.

Chapter 3

EMPLOYMENT CREATION AND REPLACEMENT AT THE NATIONAL LEVEL

3.1. MODELING THE U.S. ECONOMY

3.1.1. The IMPLAN Inter-Industry Model

Transactions between many industries are required in the production of the equipment used in transportation, as well as in providing the fuels that are the source of the energy. These transactions spread throughout the economy as the directly affected industries alter their transactions with other industries with which they deal, which in turn affect still other industries. These effects can be estimated using input-output modeling, which is described in detail in Appendix 4. As discussed previously, the present report estimates, the employment impacts of the transformation to a hydrogen economy with the use of a 509 industrial sector IMPLAN, input-output model of the U.S. economy, a 509-sector input-output model of the U.S. economy in 2002.[1] The IMPLAN model is run for an essentially all-gasoline economy and for the two market penetration scenarios of a hydrogen economy described in section 2. The employment impacts of the transformation of the economy to hydrogen are the differences in employment in various industries between the with- and without-hydrogen cases.

3.1.2. Future of the U.S. Economy in the Absence of a Hydrogen Transformation

The industries likely to be affected directly by the market expansion of hydrogen between 2020 and 2050 employed 3.3%, or 4.4 million, of the total U.S. employment of 132 million in 2005. Employment in the light vehicle manufacturing industries has decreased in the past decade, with the losses concentrated in assembly and components, while employment in vehicle sales, maintenance, and repair increased. Employment in petroleum refining has been falling in the past decade, as that industry has become more capital intensive.[2] In the base case, employment in these industries continues to evolve without the changes brought about by hydrogen market expansion.

Using the IMPLAN model, a base case for the U.S. economy was estimated. The base case depicts changes in the U.S. economy that are expected to occur in coming decades in the absence of a hydrogen transformation. The base case relies on reputable projections that have been done by others. It corresponds closely to projections by Regional Economic Models, Inc. (REMI).[3] The REMI model and its projections have been widely used for policy impact analysis by more than 50 federal and state agencies, as well as by many private firms and universities.[4] However, REMI has only 70 sectors and IMPLAN allows a user to develop a model using 509 sectors and also permits the user to modify the model's internal tables which characterize the structure of production, which was of central importance to this study. Both of these features rendered IMPLAN more suitable for the employment analysis of this report.

3.1.3. Modification of IMPLAN for Hydrogen Economy Scenarios

The unmodified version of IMPLAN used for the base case is satisfactory for modeling the future in the absence of a hydrogen transformation. However, no existing input-output model allows for widespread hydrogen adoption. Businesses that provide hydrogen-powered vehicles, stationary and mobile equipment, hydrogen infrastructure and hydrogen fuel are embedded in non-hydrogen sectors.

To remedy this limitation, the IMPLAN model was reconfigured for the present analysis for use in the hydrogen economy simulations. In particular, to characterize the changes in production caused by the market transformation to hydrogen, the purchase patterns of motor vehicle assembly, motor vehicle parts manufacturing, and industrial gases were modified with information derived from the cost data reported in Appendices 2 and 3. The procedures are reported in Appendix 4. In addition to characterize the changes in infrastructure required to produce and deliver the new products—vehicles and fuel—the industry composition of investment was altered to represent additional purchases from industries manufacturing equipment installed in these facilities as well as purchases from the construction sectors that assemble them. These changes are also reported in detail in Appendix 4.

3.2. Net Effects on Total U.S. Employment

Table 3.1 shows cumulative effects on U.S. employment in 2020, 2035 and 2050 resulting from the HFI and the Less Aggressive Scenarios, as calculated using the modeling procedures described in the preceding section. Base-case employment for each of the 509 industry sectors was subtracted from employment under each of the hydrogen adoption scenarios to obtain changes in sector employment resulting from the scenario. The results were summed to obtain Table 3.1. The net effect is the effect on total number of people at work in the U.S. economy resulting from gains in some industries and losses in others due to the hydrogen transformation. The top of the table gives numbers of workers affected. The bottom of the table gives the numbers as a percent of total U.S. employment.

Under both the HFI and the Less Aggressive Scenarios, the net effect on U.S. employment is a slight increase to 2020, by 0.13% under the HFI Scenario and by 0.04% under the Less Aggressive Scenario. Growth of jobs in the production of hydrogen vehicles and other hydrogen-using equipment will be offset by job declines in traditional activities. Investments to create the capital goods needed in producing the equipment may have some jobs effects but that impact is likely to be negligible. Depreciation of equipment used in traditional production accompanied by its gradual replacement with hydrogen economy equipment is likely to take care of much, but not all, new investments needed by 2020, without net job creation in connection with the new investments. For some types of equipment, anticipatory investment in new infrastructure and production facilities, in preparation for expected near-term demand increases beyond 2020, can be expected to increase jobs in both construction and durable goods manufacture. The new job effects become greater in the years following 2020, due to the later more rapid pace of market expansion. By 2035, the net effect on U.S. employment is increased by 0.42% under the HFI Scenario and by 0.11% under the Less Aggressive Scenario. By 2050, under the HFI Scenario, the transformational adjustments are fully completed, with no more anticipatory investment; U.S. employment is increased by a net of 0.37%, or 675,000 jobs out of a total projected base-case employment of 184 million. Under the Less Aggressive scenario, the transformation is not fully completed by 2050 and U.S. employment is increased by a net of 0.20%, or 361,000. The difference in employment between the HFI Scenario and the base case, in 2050, is equivalent to roughly half of a year's growth of the total U.S. employment of 1.38 million, occurring at that time.

The maximum net effect in Table 3.1 is less than ½ of 1% of U.S. employment. Table 3.1 reports cumulative impacts occurring over a period of years. The annual changes leading to the cumulative impacts would be even smaller. The smallness of the impacts indicates that significant macroeconomic effects, such as large-scale unemployment or inflation, will not occur.

Table 3.1. U.S. Cumulative Gains and Losses from Shifts of Employment Between Sectors[a]

Scenario		2020	2035	2050
Numbers of Workers				
HFI	**Net Effect**	**182,840**	**677,070**	**674,500**
	Gains	252,040	754,030	751,060
	Losses	*69,200*	*76,960*	*76,560*
Less Aggressive	**Net Effect**	**58,010**	**184,560**	**360,740**
	Gains	126,680	242,820	417,390
	Losses	*68,670*	*58,260*	*56,650*
Percentage Effects on Total Employment				
HFI	**Net Effect**	**0.13%**	**0.42%**	**0.37%**
	Gains	0.17%	0.46%	0.41%
	Losses	*0.05%*	*0.05%*	*0.04%*
Less Aggressive	**Net Effect**	**0.04%**	**0.11%**	**0.20%**
	Gains	0.09%	0.15%	0.23%
	Losses	*0.05%*	*0.04%*	*0.03%*

[a]All numbers represent differences between the hydrogen scenario and the baseline. For example, there are 182,840 more workers in all sectors in 2020 according to the HFI Scenario than there would have been in 2020 according to the baseline scenario.

3.3. Job Creation and Replacement

3.3.1. Employment Shifts between Gaining and Losing Industries

Underlying the percentage changes in total numbers of people employed in the U.S. will be changes in how the work force is deployed among industries. New jobs will be created in hydrogen-related manufacturing and services industries, and these new jobs will replace jobs in non-hydrogen-related manufacturing and services industries. The IMPLAN model simulations shed partial light on job creation and replacement effects by indicating how employment will expand in some industries and contract in others.

The shifts of employment between sectors are very small as a percent of total U.S. employment. Employment shifts many-fold greater than those in Table 3.1 occur continually in the U.S. economy for reasons apart from hydrogen. Among the reasons are shifts in consumer demands for various products, industrial technological changes connected with diverse developments that occur in a growing economy, effects of environmental and other regulations, and global economic changes affecting the U.S. As with the net U.S. employment effects, the aggregate effects of job gains and losses from employment shifts between industries due to hydrogen transformation are too small to raise macroeconomic concerns.

3.3.2. Job Creation and Replacement within Key Industries

The IMPLAN sector employment changes in Table 3.1 do not take account of changes in the composition of employment within sectors. Within sectors that are key to hydrogen systems, demands for skills in hydrogen-related production will increase, and demands for skills in traditional production will decrease.

For example, within automotive sectors, in switching from producing gasoline-powered to hydrogen-powered vehicles, skills used in hydrogen-oriented production will experience increased demand, while skills used in non-hydrogen-oriented production will become outmoded. While automobile assembly primarily entails a switch from producing one kind of vehicle to another with no expected effect on labor productivity in the sector and hence no effect on employment in the sector, a part of the sector's employment is subject to skill change. Even though much assembly may continue as is, some specialized assembly tasks are likely to be required for fuel cells and new types of powertrains requiring new blue collar skills. At the engineering level, not as many mechanical and other engineers will be needed, but electrical, electronics and chemical engineers will be in greater demand. As another example, in the auto repair sector the total number of service technicians is not estimated to be greatly affected by the switch to hydrogen. However, service technicians in the future will need to have expertise to service hydrogen fuel cell vehicles. As the transformation progresses, service technicians who only service internal combustion vehicles will find their skills in reduced demand if they are unable also to service hydrogen vehicles. Ultimately, substantial turnover in skills will be required. Experience with hybrid vehicle technologies will form the basis for many of these skill changes.

The foregoing examples are for sectors where employment is not much affected, but skill composition changes are large. As a contrasting case, industrial gas manufacturing, with its large increase in hydrogen production, will undergo large employment changes due to industry demand growth even though the skill composition within the industry does not change much. The fuel cells and batteries sector experiences a similar change driven by demand growth. The employment changes for these sectors are fully included in the employment shifts between industries in Table 3.1. While small as part of the national effect, the employment changes for key hydrogen sectors may be important for considering education, training and re-training needs in these sectors.

Table 3.2 (white collar workers) and Table 3.3 (blue collar workers) present estimates of cumulative job creation and job replacement within several industrial sectors that will be key to the expansion of hydrogen markets. "Job creation" connected with changes in skill composition within an industry is defined as the addition of a job whose skill requirements cannot be met by workers with the skills of presently employed workers without outside re-tooling of skills. "Job replacement" is the loss of a job of a worker who is not qualified to fill a newly created job vacancy without such training. Tables 3.2 and 3.3 pertain to a part of the employment in a sector, either white collar or blue collar, and are congruent with the total sector employment changes from the IMPLAN model. The job creation and job replacement figures within sectors shown in Tables 3.2 and 3.3 are medium term estimates based on opinions of people with first-hand knowledge of the industries and on analysis of data on occupational breakdowns of employment in each industry from the Bureau of Labor Statistics (BLS).[5] The occupational breakdown is very detailed, with figures for over 200 occupations in some industries. Appendix 5 contains a complete description of how the occupational and IMPLAN modeling results were used to produce the estimates of creation and replacement by occupation shown in Tables 3.2 and 3.3.

Table 3.2. Cumulative White Collar Job Creation and Replacement within Sectors[a]

Industry	HFI			Less Aggressive		
	2020	2035	2050	2020	2035	2050
Auto Assembly						
Net Change	-	-	-	-	-	-
Creation	580	1,770	1,840	20	300	1,100
Replacement	*580*	*1,770*	*1,840*	*20*	*300*	*1,100*
Auto Body Mfg						
Net Change	-	-	-	-	-	-
Creation	100	330	360	5	60	210
Replacement	*100*	*330*	*360*	*5*	*60*	*210*
Auto Parts Mfg, incl. Fuel Cells						
Net Change	-	-	-	-	-	-
Creation	3,220	10,640	12,060	100	1,780	7,240
Replacement	*3,220*	*10,640*	*12,060*	*100*	*1,780*	*7,240*
Hydrogen Production						
Net Change	600	8,510	10,050	120	1,100	4,070
Creation	600	8,510	10,050	120	1,100	4,070
Replacement	*0*	*0*	*0*	*0*	*0*	*0*
Refining						
Net Change	-30	-370	-400	-10	-100	-210
Creation	0	0	0	0	0	0
Replacement	*30*	*370*	*400*	*10*	*100*	*210*
Construction						
Net Change	1,550	1,470	1,340	1,040	980	900
Creation	1,790	15,250	25,940	1,040	2,280	15,540
Replacement	*240*	*13,780*	*24,600*	*0*	*1,300*	*14,640*

[a] All numbers are cumulative full-time equivalent jobs (FTEs) from 2018. Job replacement is the elimination of jobs, and net change is creation minus replacement. For example, there are 8,510 more white collar jobs in the hydrogen production sector in 2035 under the HFI than there would have been under the baseline scenario. Those jobs are created starting in 2018, meaning the 600 jobs created in that sector from 2018 to 2020 are included in the 8,510.

Table 3.3. Cumulative Blue Collar Job Creation and Replacement within Sectors[a]

Industry	HFI			Less Aggressive		
	2020	2035	2050	2020	2035	2050
Auto Assembly						
Net Change	-	-	-	-	-	-
Creation	3,150	9,580	9,950	100	1,610	5,970
Replacement	*3,150*	*9,580*	*9,950*	*100*	*1,610*	*5,970*
Auto Body Mfg						
Net Change	-	-	-	-	-	-
Creation	1,560	4,970	5,440	50	830	3,270
Replacement	*1,560*	*4,970*	*5,440*	*50*	*830*	*3,270*
Auto Parts Mfg, incl. Fuel Cells						
Net Change	-	-	-	-	-	-
Creation	31,280	103,560	117,360	940	17,360	70,410
Replacement	*31,280*	*103,560*	*117,360*	*940*	*17,360*	*70,410*
Auto Dealerships & Repair						
Net Change	-	-	-	-	-	-
Creation	25,130	435,770	681,730	760	50,510	409,040
Replacement	*25,130*	*435,770*	*681,730*	*760*	*50,510*	*409,040*
Hydrogen Production						
Net Change	2,450	34,550	40,780	480	4,480	16,540
Creation	2,450	34,550	40,780	480	4,480	16,540
Replacement	*0*	*0*	*0*	*0*	*0*	*0*
Refining						
Net Change	-180	-2,300	-2,480	-40	-650	-1,290
Creation	0	0	0	0	0	0
Replacement	*180*	*2,300*	*2,480*	*40*	*650*	*1,290*
Construction						
Net Change	3,880	3,700	3,360	2,600	2,470	2,250
Creation	5,460	50,180	85,670	2,600	7,430	51,460
Replacement	*1,580*	*46,480*	*82,310*	*0*	*4,960*	*49,210*

[a] All numbers are cumulative, full-time equivalent jobs (FTEs) from 2018. Job replacement is the elimination of jobs, and net change is creation minus replacement. For example there are 34,550 more blue collar jobs in the hydrogen production sector in 2035 under the HFI than there would have been under the baseline scenario. Those jobs are created starting in 2018, meaning the 2,450 jobs created in that sector from 2018 to 2020 are included in the 34,550.

As expected, the cumulative effects in these industries become greater over time, and they are greater for the HFI than the Less Aggressive Scenario. Common influences for the two hydrogen scenarios are as follows.

Automotive Sectors. For the automotive sectors (vehicle assembly, body manufacturing, parts manufacturing, and repairs), employment creation and replacement within each sector is the dominant influence. Total employment in these sectors is not forecast to change appreciably, in view of the expectation that total labor productivity in automobile production will not be significantly affected by the switch from gasoline to hydrogen vehicles. Some tasks within the automotive sectors will remain the same, such as task involved in producing and assembling certain automobile parts e.g., automobile wheels. But new skills will be required in other tasks, such as producing and installing fuel cells new powertrains and other associated equipment. These new jobs will utilize workers with experience and training different from that possessed by workers in the production of gasoline-powered vehicles.

Hydrogen Production. Important changes in types of jobs within the hydrogen production sector are not anticipated because the majority of the hydrogen production process is expected to be similar to the industrial gas processes that exist today. One notable difference is introduced with coal gasification, with its use of solid inputs. The use of solid inputs would change the occupational structure by adding staff for coal handling at the hydrogen production sites, similar to the coal-handling performed by employees at today's coal-fired power plants. However, these inventories are handled by small numbers of workers whose addition should be noted but would not affect the occupational structure of the sector significantly.[6] In this sector, the dominant influence is job creation for all skill levels to meet the increasing demand for hydrogen.

Petroleum Refining. Skill sets in the petroleum refining sector are not changed by the market transformation. However, jobs will be lost due to the emergence and substitution of hydrogen-based fuels. Consequently, the change in employment in this sector reflects employment losses in Table 3.1. The petroleum refining employment losses, resulting from the reduction in the demand for gasoline, are further detailed in Tables 3.2 and 3.3.

Construction. Increased pipeline construction for carrying hydrogen fuel from centralized production sites to dispensing and other use sites will lead to increased employment in the construction sector and will create jobs for all workers in pipeline construction. The increasing importance of pipeline construction relative to industrial buildings will change the skill mix in this sector. Job creation and replacement in this sector thus represent a combination of increases in sector employment and changes in skill mix.

3.4. Education, Training, and Re-Training

3.4.1. Human Capital in the Hydrogen Market Expansion

A Variety of Needs. Education, training and re-training needs will depend on the skills involved in particular jobs. Some jobs requiring university education will call for re-direction in the capacity and nature of engineering programs offered in universities. Assembly line workers accustomed to routine tasks in the manufacture and installation of gasoline engines may be able to turn to hydrogen powertrain tasks with only a minimum of on-the-job instruction or in-house re-training. For high-skill blue collar workers, technical differences in basic technology used in gasoline- and hydrogen-powered vehicles, accentuated by the increasingly skill-intensive use of computers in automobiles, will change the content of vocational education. Entering workers will receive instruction with new content. Those who began their careers before the advent of hydrogen vehicles will need vocational re-training. In addition to workers involved in producing vehicles, this group includes vehicle service technicians and workers involved in delivering and dispensing fuel to automobiles.

Short-Term vs. Long-Term Adjustment. Rapid rates of job creation in the early years of hydrogen market expansion may lead to bottlenecks in the availability of workers for jobs being created, with the growth in number of workers preparing for newly created jobs lagging behind job openings. The more rapid the job creation, the more likely it is that growth in job availability will be under-estimated at the time career decisions are being made, and the more likely that there will be difficulties in finding qualified teachers and in gearing up for expanded education and training programs in new fields. For cases where small numbers of workers are involved, the changes that need to be made in university and vocational education may be limited. For occupations where a large number of workers relative to present numbers are needed and the skill levels require highly technical instruction, lags in meeting instructional needs could occur that might interfere with, or at least make difficult, the transformation to a hydrogen economy. In some cases, even though numbers are small, the new skill sets required for the hydrogen economy may be so different, so specialized and so highly education-intensive that bottlenecks may develop in supplying enough specialists to meet demands.

Over time, much of the response to job creation and replacement will come about through retirement of older workers and the education and training of younger, entering workers. If the number of jobs being replaced is smaller than the number of workers in those jobs who are retiring, and if entering workers are preparing themselves for the new jobs being created, few noticeable education, training, or re-training concerns may arise. Given that a worker's career spans approximately 40 years, on the order of 2½% of workers in an occupation will retire in any one year.[7] An annual rate of job replacement of 2½% provides an approximate dividing line between whether replacement will exceed or fall short of retirement. As an example of the variation in this rate that could be caused by demographic differences, the baby boomers, who are expected to retire between 2011 and 2029, would retire at a rate of about 2.7% per year, as opposed to the 2.5% assumed above. More rapid changes could mean that jobs are replaced faster than older workers retire—resulting in an excess of workers wishing to remain in these jobs, who then turn to other jobs with or without re-training, become unemployed, or retire early. For re-training to enable displaced workers to complete a normal career path, a series of events must transpire: a re-training course must be offered, the displaced workers must be willing to enroll in the course, they must complete the course, and they must find jobs afterwards. Because of slippages in this process, some unemployment and premature exit from the labor force could occur.

3.4.2. White Collar Workers

Occupational Composition. White collar workers include engineers, engineering managers, drafters, and engineering technicians. Engineers and engineering managers comprise the majority of this group and have university-level education. Drafters and engineering technicians receive training mostly from vocational and technical schools.

Industry Opinions on Skill Changes. In the interviews of persons with first-hand knowledge of developments in the automobile industry, referred to in the preceding section, most interviewees believed that the current set of engineering skills used for conventional internal combustion engine vehicles will likely change significantly. They agreed mechanical engineering skills will focus less on purely mechanical functions and more on developing electro-mechanical systems. Some suggested that this will be particularly true in the event that alternative platforms such as by-wire technology (in which steering, braking, and acceleration will be controlled electronically rather than mechanically) are adopted. Others suggested that the traditional vehicle platform will evolve but not change radically. On balance, it appears that new designs will be required for hydrogen-powered vehicles but non-propulsion-related systems such as heating and cooling systems will likely employ the same mechanical engineering skills that as are in use today.

Regardless of design, interviewees believed that changes in required engineering skills will be evolutionary rather than revolutionary. Examples of evolutionary training changes in the auto industry include the introductions of the front-wheel drive system and various fuel injection systems, and the use of different on-board computer systems. In all of these cases, the automotive industry will rely upon internal re-training and on-the-job experience, as well as changes in curricula at universities and community colleges to support the turnover in new skills.

Most interviewees agreed that as a result of work on hybrid electric vehicles, there are already a significant number of engineers with either educational backgrounds or on-the-job experience in electronics, controls, sensors, control system integration, power controllers, electric motors, on-board computing, and system integration. New elements required in a fuel cell vehicle system include the fuel stack and some attendant sensing and control systems. But a great many system requirements are already present in conventional and hybrid vehicle systems.

Estimation of Education, Training, and Re-Training Needs. The cumulative effects that were discussed in connection with Tables 3.1, 3.2 and 3.3 are estimates of how the future industrial structure of the economy will be affected by the hydrogen transformation. However, in addition to studying the cumulative impacts of the hydrogen transformation, it is important to consider how the impacts in any given year will determine the number of additional people that will need to complete education and training to equip them for jobs created that year and the number of additional jobs that will be replaced in that year, thereby requiring layoffs in that year for workers who are not re-trained. Tables 3.4 and 3.6 provide an analysis of the job impacts on white collar and blue collar workers for key industries on an annual basis for three representative years- 2020, 2035, and 2050. The results show how many new jobs are expected to be created in those years and how many old jobs disappear in those years. In the tables, the average annual job changes are expressed as a percentage of the number of workers in seven affected occupations. These percentages facilitate discussion of the burdens that will be placed on education and training relative to existing numbers receiving instruction in the case of job creation, and comparison with expected industry retirement reates in the case of job replacement. The first step in deriving Table 3.4 is to calculate the cumulative percentage impacts on white collar employment in each industry from the beginning of market expansion in 2018 up to 2020, 2035, or 2050, as appropriate. These percentage impacts are obtained by dividing the cumulative effect on number of white collar workers given in Table 3.2 by the total number of white collar workers in affected occupations in the industry. Because the impacts will be spread over 2 years for 2020, 17 years for 2035, and 32 years for 2050, the annual figures in Table 3.4 are obtained by dividing the cumulative percentage impacts by the appropriate number of years.

Under the HFI Scenario, the highest positive average annual percentage impacts on job creation in Table 3.4 are 10.41% for hydrogen production job creation in 2035 and 4.66% for auto parts (including fuel cells) in 2020.

Most of the data presented in Table 3.4 involves situations where job replacement is as large or nearly as large as job creation. The average annual percentage impacts are generally less than 2 ½% and most often are far less. As a point of comparison, churning rates, a measure of job creation and replacement under normal conditions, defined in footnote 32, are available for many industries in the economy. Typical churning rates are 13% to 28% of an industry's employment; those of vehicle components and vehicle body manufacturing have been 7% and 9% recently, with considerable variation among states.[8] The much smaller percentages in Table 3.4 indicate that employment impacts from job creation and replacement brought on by hydrogen transformation are small compared to the normal employment churning that occurs in the economy. Furthermore, because many of the annual percentage rates of replacement will be substantially less than 2½% it is likely that retirement rates should be sufficient to adjust to declining demands for outmoded jobs, generally without downward employment pressures on workers not choosing to retire

Because the number of engineers in each discipline working in the hydrogen area is small compared to the total number of engineers working in the U.S., the current education system easily will have the capacity to train engineers needed in the hydrogen economy over the longer run.

Table 3.4. Average Yearly Percentage Impacts, White Collar Employment[a]

Industry	HFI			Less Aggressive		
	2020	2035	2050	2020	2035	2050
Auto Assembly						
Net Change	-	-	-	-	-	-
Creation	3.96%	1.56%	0.93%	0.12%	0.26%	0.56%
Replacement	*3.96%*	*1.56%*	*0.93%*	*0.12%*	*0.26%*	*0.56%*
Auto Body Mfg						
Net Change	-	-	-	-	-	-
Creation	2.66%	1.05%	0.63%	0.08%	0.18%	0.38%
Replacement	*2.66%*	*1.05%*	*0.63%*	*0.08%*	*0.18%*	*0.38%*
Auto Parts Mfg, incl. Fuel Cells						
Net Change	-	-	-	-	-	-
Creation	4.66%	1.83%	1.10%	0.14%	0.31%	0.66%
Replacement	*4.66%*	*1.83%*	*1.10%*	*0.14%*	*0.31%*	*0.66%*
Hydrogen Production						
Net Change	6.11%	10.41%	6.23%	1.19%	1.35%	2.53%
Creation	6.11%	10.41%	6.23%	1.19%	1.35%	2.53%
Replacement	*0.00%*	*0.00%*	*0.00%*	*0.00%*	*0.00%*	*0.00%*
Refining						
Net Change	-0.36%	-0.66%	-0.49%	-0.08%	-0.19%	-0.25%
Creation	0.00%	0.00%	0.00%	0.00%	0.00%	0.00%
Replacement	*0.36%*	*0.66%*	*0.49%*	*0.08%*	*0.19%*	*0.25%*
Construction						
Net Change	1.32%	0.13%	0.06%	0.88%	0.09%	0.04%
Creation	1.52%	0.95%	1.11%	0.88%	0.20%	0.67%
Replacement	*0.21%*	*0.82%*	*1.06%*	*0.00%*	*0.11%*	*0.63%*

[a] All averages are from 2018 to the year listed. For example, in auto assembly, on average, 3.96% of skill-affected white collar jobs are created and replaced each year from 2018 to 2020, 1.56% are created and replaced each year from 2018 to 2035, and 0.93% are created and replaced each year from 2018 to 2050.

The possibility of shorter run problems remains to be considered. Table 3.5 compares the total number of engineers in the U.S. to the number of engineers in key industrial sectors. Numbers are presented for the major engineering disciplines that are most likely to be involved in the transformation to a hydrogen economy.

Table 3.5 indicates that in the hydrogen economy, engineers will still need to complete basic undergraduate programs at a minimum, but will need to include hydrogen-related subjects in their undergraduate curriculum. As undergraduate education curricula develop and as demand for engineers in the fuel cell and hydrogen technologies areas increases, there should be sufficient numbers of students to fill the need for qualified engineers. However, it takes approximately 4 to 8 years for engineers to complete professional education, including undergraduate and graduate schools. The high annual growth rates in the hydrogen production and fuel cell sectors to 2020 suggest that shortages of engineers trained for these sectors could arise in the early years of vigorous hydrogen market expansion, although it should be kept in mind that the absolute number of engineers in these industries is small.

The engineering disciplines needed to develop technology solutions for hydrogen systems include chemical engineering, electrical/electronics engineering, industrial engineering, mechanical engineering, and materials engineering. R&D activity at universities increases the number of personnel and the knowledge base for speeding up the educational response. Development of educational programs and courses at universities has begun, and much progress is apparent. However, very few degree programs specific to fuel cells and hydrogen technology are in place. Currently, approximately 60 colleges and universities offer research and coursework in fuel cells and hydrogen technology. Four schools in the U.S. have been identified that have degree programs specifically targeted to fuel cells or alternative energy technologies.[9] Most of the activity is at research centers. At least 20 schools have established multidisciplinary groups and research centers, many of which include industry partners.

Re-training needs for white collar workers, both engineers and non-engineers, will depend on the pace of new technological developments as the hydrogen transformation proceeds. Continuing education courses, workshops, and on-the-job training are likely means of carrying out the white collar re-training.

3.4.3. Blue Collar Workers

Occupational Composition. Blue collar workers include, but are not limited to, manufacturing employees, construction employees, automotive service and repair technicians, service station attendants, and hydrogen fuel deliverers. Blue collar workers obtain training from vocational and technical schools where necessary.

Table 3.5. Comparison of Total Engineers Nationwide to Engineers in Sectors Directly Affected by Hydrogen ("Key" Sectors), by Engineering Discipline

Engineering Occupations	Total U.S., 2005	Total Key Sectors, 2005	% in Key Sectors
Chemical Engineers	27,280	1,415	5%
Civil Engineers	226,900	2,608	1%
Electrical/Electronics Engineers	313,040	1,757	1%
Health and Safety Engineers	24,110	1,058	4%
Industrial Engineers	185,620	18,702	10%
Mechanical Engineers	210,790	12,180	6%
Materials Engineers	20,850	442	2%
Total	1,008,590	35,912	4%

Source: Bureau of Labor Statistics (BLS), Occupational Employment Statistics, data as of May 2005, accessed August-September 2006 at http://data.bls.gov/oes/search.jsp?data_tool=OES.

Industry Opinions on Skill Changes. Interviewees generally agreed that blue collar jobs in manufacturing will be significantly affected by a transformation to hydrogen fuel cell vehicles. Most interviewees focused on the level of training of workers on the line. Computer literacy skills, knowledge of electrical systems, and the ability to use computerized diagnostic equipment will be particularly important for assembly skills in the future. Machining skills related to internal combustion engine construction are not likely to be needed, as fuel cell stacks require little to no machining and are unlikely to be manufactured by the automobile companies themselves. Skills related to assembly of electro-mechanical systems and computer hardware would be in greater need. The balance of components required to complete the fuel cell system requires skills for assembly similar to those for a traditional gasoline engine.

As with engineering skill development, interviewees believed that training will be evolutionary rather than revolutionary. In addition to training courses on skills needed for increasingly computerized vehicles and manufacturing techniques, on-the-job training will assist in the transformation. The development of hybrid vehicles has provided an environment for the development of new manufacturing skills. For example, dedicated lines for hybrid vehicle systems have provided much-needed learning about procedures, failure rates, stoppage points, line coordination, and other matters. The dedicated lines thus were in some respects the classroom that was used to train the production in the supporting industries, and interviewers indicated that the same sort of experience likely would emerge in the production of hydrogen fuel cell vehicles.

In the area of vehicle service and repair, all interviewees agreed that the most significant skill set change will be in troubleshooting, repair, and service of propulsion systems. They believe that these skills will be very new, and very technical in nature. Technicians will have to become more competent in computer and electrical system maintenance. However modularization of key components such as batteries, fuel cell stacks, and power converters, may reduce the amount of re-training needed. Similar to production, hybrid vehicle maintenance will provide a means of developing new repair and maintenance skills. However, hydrogen fuel cell vehicles will require new safety training at all levels because of their use of hydrogen and high-voltage electrical systems.

Estimation of Training and Re-training Needs. Table 3.6 shows average annual percentage impacts on the blue collar workers for industry sectors of interest. The percentage impacts on blue collar workers are very similar to the white collar impacts in Table 3.4. The Automobile Dealer and Repair sector, excluded in Table 3.4, is added in Table 3.6. While appreciable changes in white collar workers are not expected in this sector, considerable impact is expected for the major blue collar category, auto service technicians. The same general comments apply to Table 3.6 as were made for Table 3.4. These are the rapid increase required in auto parts (including fuel cells) manufacture and hydrogen production in the early years of a rapid transformation, the otherwise small impacts compared to normal employment churning in the economy, and the excess of worker retirements over job replacement rates.

The training and re-training implications for blue collar workers are different from the white collar implications. Many more blue collar workers are affected, because the numbers of blue collar workers are many-fold greater than those of white collar workers. Examples of blue collar workers in the key sectors include employees involved in manufacturing, construction, and repair and maintenance of both vehicles and production facilities. Training for the blue collar group will depend on the employee functions and may range from as little as basic safety training to very specialized automotive technician training.

The employment impacts are modeled on annual average levels, as full-time equivalents. Because not all workers are year-round workers, particularly among the blue collar labor force, more people have some work experience during a typical year than is indicated by the full-time equivalent calculation. Consideration of all persons with work experience during the year by industry sector could increase the retraining needs identified below to a modest extent, more so in some sectors than others. For example, nearly 96% of durable manufacturing workers, which includes the auto sectors were full-time in 2005, and 94% of nondurable manufacturing workers, which include refining and industrial gases, were full time. Full-time employment was lower in construction, around 90% in construction that includes residential and commercial buildings as well as pipelines; 70% of full-time construction workers in all construction sectors worked 50 to 52 weeks that year.[10]

Manufacturing and Construction. Much of the training and re-training for automobile manufacturing and construction workers as a result of hydrogen technologies may be accomplished through on-the-job training sponsored by employers. In the manufacturing area, for example, there are large numbers of assemblers and machine operators who may be affected as production lines are changed to accommodate potential changes in vehicle designs. In the construction area laborers may need safety training for working with hydrogen pipelines, while plumbers, pipe fitters and welders may need training on new piping specifications and safety requirements. These few examples are not inclusive of all affected manufacturing and construction occupations, but are meant to illustrate the wide range of blue collar workers who may be affected by the introduction of hydrogen technologies.

Automotive Service Technicians and Mechanics. Automotive service technicians and mechanics are a large group which will be significantly affected by a change to hydrogen fuel cell vehicles. Automotive service technicians and mechanics receive training at vocational and technical schools, and most take standard certification tests in order to be qualified to work on vehicles. According to the Bureau of Labor Statistics, as of May 2005, there were 654,000 automotive service technicians and mechanics in the U.S. working in dealerships and independent maintenance shops; additional workers are employed at large firms with extensive light-duty vehicle fleets in other industries.[11] This figure does not include automotive body-related repairers, automotive glass installers, or tire changers and repairers. The study uses data from the BLS Quarterly Census of Employment and Wages (QCEW) Program, which does not cover self-employed individuals and agricultural workers who are not subject to unemployment insurance payroll taxes. Accordingly, the prevalence of self-employed individuals in the automotive service and repair industry could tend to increase re-training needs among those skill sets somewhat beyond the extent estimated here.

Table. 3.6. Average Yearly Percentage Impacts, Blue Collar Employment[a]

Industry	HFI			Less Aggressive		
	2020	2035	2050	2020	2035	2050
Auto Assembly						
Net Change	-	-	-	-	-	-
Creation	2.00%	0.78%	0.47%	0.06%	0.13%	0.28%
Replacement	*2.00%*	*0.78%*	*0.47%*	*0.06%*	*0.13%*	*0.28%*
Auto Body Mfg						
Net Change	-	-	-	-	-	-
Creation	2.00%	0.78%	0.47%	0.06%	0.13%	0.28%
Replacement	*2.00%*	*0.78%*	*0.47%*	*0.06%*	*0.13%*	*0.28%*
Auto Parts Mfg, incl. Fuel Cells						
Net Change	-	-	-	-	-	-
Creation	3.99%	1.57%	0.94%	0.12%	0.26%	0.56%
Replacement	*3.99%*	*1.57%*	*0.94%*	*0.12%*	*0.26%*	*0.56%*
Auto Dealerships & Repair						
Net Change	-	-	-	-	-	-
Creation	1.00%	1.46%	1.81%	0.03%	0.24%	1.09%
Replacement	*1.00%*	*1.46%*	*1.81%*	*0.03%*	*0.24%*	*1.09%*
Hydrogen Production						
Net Change	6.11%	10.41%	6.23%	1.19%	1.35%	2.53%
Creation	6.11%	10.41%	6.23%	1.19%	1.35%	2.53%
Replacement	*0.00%*	*0.00%*	*0.00%*	*0.00%*	*0.00%*	*0.00%*
Refining						
Net Change	-0.36%	-0.66%	-0.49%	-0.08%	-0.19%	-0.25%
Creation	0.00%	0.00%	0.00%	0.00%	0.00%	0.00%
Replacement	*0.36%*	*0.66%*	*0.49%*	*0.08%*	*0.19%*	*0.25%*
Construction						
Net Change	0.57%	0.06%	0.02%	0.38%	0.04%	0.02%
Creation	0.79%	0.53%	0.63%	0.38%	0.11%	0.38%
Replacement	*0.23%*	*0.48%*	*0.60%*	*0.00%*	*0.07%*	*0.36%*

[a] All averages are from 2018 to the year listed. For example, in auto assembly, on average, 2.00% of skill-affected blue collar jobs are created and replaced each year from 2018 to 2020, 0.78% are created and replaced each year from 2018 to 2035, and 0.47% are created and replaced each year from 2018 to 2050.

Two organizations are responsible for developing and certifying training programs for automotive service technicians and will have to be involved in program changes: the National Automotive Technicians Education Foundation (NATEF), and the National Institute for Automotive Service Excellence (ASE). The NATEF is an independent, non-profit organization which evaluates technician training programs against standards developed by the automotive industry and recommends qualifying programs for certification by ASE. NATEF also evaluates the providers of in-service technician training programs under its Continuing Automotive Service Education (CASE) program. The ASE is an independent, non-profit organization that provides testing and certification of repair and service professionals.

Currently, automotive vocational and technical training based on gasoline internal combustion engines in the U.S. is offered at approximately 2,100 schools. Most programs take 1 to 2 years to complete. As an example of programs offered, the Universal Technical Institute (UTI), with operations nationwide, provides accredited gasoline automobile programs in auto service and repair. These programs include a basic program that lasts a minimum of 51 weeks. Enrollees can gain more specific skills by taking manufacturer -specific certification programs that are approximately 16 weeks long (the manufacturer -specific certification programs offered at UTI include Toyota/Lexus, and Ford/Lincoln/Mercury). At the most advanced level, students can take additional manufacturer-specific programs (Audi, BMW, Mercedes, Volkswagen, Volvo) that last approximately 20 weeks. These manufacturer-specific programs indicate that there is a high level of coordination between the servicing industry and these vocational schools.

In the future, to meet demands for servicing of the hydrogen vehicle fleet drawing on training and re-training at these schools, the growth in numbers of qualified service personnel must exceed that of vehicle sales. Vehicle buyers will not be willing to purchase hydrogen vehicles unless they are assured of the availability of competent and conveniently located service allowing flexibility in where they choose to drive. As many as 20% of mechanics and technicians, or approximately 110,000, could need hydrogen-oriented training to be prepared to service the hydrogen vehicles that are estimated to be on the road by 2020 under the HFI Scenario. By 2035, 60% of the stock of total vehicles will be hydrogen vehicles under the HFI Scenario, requiring approximately 340,000 trained mechanics and technicians to service them. By 2050, when the vehicle stock is 95% hydrogen under the HFI Scenario, approximately 630,000 would need to know how to service hydrogen vehicles. These figures are summarized in Table 3.7.

Table 3.7. Potential Demand for Automotive Service Technicians and Mechanics Trained in Hydrogen Vehicles under the HFI Scenario[a]

Year	Total Projected Service Technicians	Demand for Hydrogen Trained Service Technicians	%
2020	539,000	110,000	20%
2035	561,000	340,000	60%
2050	663,000	630,000	95%

[a] Values estimated using IMPLAN model results and hydrogen vehicle stock penetration assumptions.

Two major types of training will be needed—initial training for workers starting their careers with no previous training and re-training for workers already trained in gasoline engine technology. Because the vehicle stock will include a significant number of gasoline vehicles for many years, overall training programs will need to address both technologies. A major emphasis on getting the new programs in place will be required to meet demands for servicing the hydrogen vehicle fleet. The effort includes coordination, and mustering resources necessary to develop, certify, and deploy training for the automotive service technicians and mechanics.

Work has started in this area but is still in the beginning stages. For example, in California, the College of the Desert and the Sun Line Transit Agency have developed the first training program for hydrogen fuel cell buses called "Hydrogen Fuel Cell Engines and Related Technologies Course Manual."[12] The National Alternative Fuel Training Consortium, headquartered at the University of West Virginia, provides a one-day course on the basics of hydrogen production and potential uses in vehicles, as well as an introductory workshop for hydrogen-powered vehicles.[13]

Fueling Stations and Hydrogen Delivery. Vocational training and re-training needs not considered above pertain to service attendants at retail fuel stations who will require some training. The importance of fuel station instructional needs is suggested by the fact that there were are an estimated 167,000 gasoline stations in the U.S. as of May 2006.[14] Similarly, truck drivers who deliver hydrogen to fueling stations will require special instruction in handling hydrogen fuel.

End Notes

[1] Minnesota IMPLAN Group, Inc., *IMPLAN Professional, Version 2.0, Social Accounting & Impact Analysis Software, User Guide, and Data Guide,* 3rd Edition Stillwater, Minn., February 2004. IMPLAN is a static economic model, in so far as it does not contain models of investment and labor supply responses. However, in light of the relative magnitude of hydrogen investments and the lack of detailed knowledge regarding the timing of hydrogen-related actions 25 to 45 years in the future, a static model is satisfactory for the purposes of this analysis. Since IMPLAN is a national rather than an international model, the investigation of international competition relies on other models.

[2] Bureau of Labor Statistics, Quarterly Census of Employment and Wages (QCEW), accessed August – September 2006 at http://www.bls.gov/cew/home.htm.

[3] Regional Economic Models, Inc., *Policy Insight Model Version 8.0* (Amherst, Mass., Spring 2006); documentation of all elements utilized in the REMI forecasts are contained in Regional Economic Models, Inc., *Model Documentation: REMI Policy Insights Version 8.0* (Amherst, Mass., 2006), pp. 51-71. REMI forecasts of component demands are tied to Bureau of Labor Statistics 10-year forecasts of the national input-output matrix, published in Berman, J.M., "Industry Output and Employment Projections to 2012," *Monthly Labor Review* 127, No. 2 (February 2006), 58-79.

[4] Accessed August – September 2006 at http://www.remi.com/support/clients.shtml.

[5] Bureau of Labor Statistics (BLS), Occupational Employment Statistics, data as of May 2005, accessed August-September 2006 at http://data.bls.gov/oes/search.jsp?data_tools=OES.

[6] Bureau of Labor Statistics, Occupational Employment Statistics: Electric Power Generation, Transmission and Distribution (NAICS code 221100), accessed August – September 2006 at http://data.bls.gov/oes/search.jsp.

[7] Social Security Administration, Office of Policy Data, Annual Statistical Supplement, 2005, Tables 4.B1, 6.A1, accessed September 2006 at http://www.ssa.gov/policy/docs/statcomps/supplement/2005/4b.pdf.

[8] Brown, C., Haltiwanger, J., Lane, J., *Economic Turbulence: Is a Volatile Economy Good for America?* (Chicago: University of Chicago Press, 2006), Table 4.1, p. 48. Churning is defined as the average of job accessions (job creation) and separations (job replacement) as a percent of total employment in the industry. Thus, churning is measured as ½ x (accessions plus separations)/total employment: Abowd, J. M., Stephens, B. E., Vilhuber, L., Anderson, F., McKinney, K.L., Roemer, M., Woodcock, S., "The LEH Infrastructure Files and the Creation of the Quarterly Workforce Indicators," Technical Paper No. TP-2006-01, U.S. Census Bureau, Silver Hill, Md., December 5, 2005, p. 108. Data on churning rates by industry and state are available at http://lehd.dsd.census.gov/led/datatools/qwiapp.html.

[9] U.S. Department of Energy, Office of Energy Efficiency and Renewable Energy, accessed August – September 2006 at http://www.eere.energy.gov/hydrogenandfuelcells/education/related_prog.html.

[10] Bureau of Labor Statistics, "Wage and salary workers with work experience in 2005 by industry of the job held the longest, March 2006," Current Population Survey, Washington, D.C., 2006.

[11] Bureau of Labor Statistics, Occupational Employment Statistics, accessed August – September 2006 at http://data.bls.gov/oes/search.jsp?data_tool=OES, search of Automotive Service Technicians and Mechanics (493023) for all sectors and all industries, as of May 2005.

[12] U.S. Department of Energy, Office of Energy Efficiency and Renewable Energy, accessed August – September 2006 at http://www.eere.energy.gov/hydrogenandfuelcells/tech_validation/h2_manual.html.

[13] The National Alternative Fuel Training Consortium, accessed August 2006 at http://www.naftc.wvu.edu/.

[14] *National Petroleum News* 2006 Survey, www.npnweb.com, annual survey of all retail outlets of any kind at which the public can buy gasoline.

Chapter 4

REGIONAL VARIATION IN ECONOMIC IMPACTS

4.1. CHARACTERISTICS OF SELECTED REGIONS

Five contrasting geographic regions illustrate the potential for differences in the impacts of a transformation to hydrogen across the nation. These regions include:

Upper Midwest
Lower New England and Upper Mid-Atlantic
California
Tennessee
Houston/Galveston

The Upper Midwest is a cluster of states consisting of Ohio, Michigan, Indiana, Illinois, and Wisconsin. This region has the greatest concentration of auto and auto parts manufacturing in the country.

The Lower New England and Upper Mid-Atlantic regions consist of Massachusetts, Rhode Island, Connecticut, New York, Pennsylvania, and New Jersey. This region has a more diversified economy. It is broadly representative of the national economy, but is distinguished by the fact that it is a large importer of oil-based products, including gasoline.

California has several unique economic characteristics. It has a substantial but declining refining and petrochemical sector; it has been under significant industrial transformation in recent years; and it has been at the forefront of energy conservation and the use of alternative energy sources.

Tennessee is representative of states that have been participating in the new part of the auto industry. Vehicle assembly and automotive parts manufacturing have represented a growing share of the state economy, stimulated by the location initially of foreign, and then domestic, automotive firms.

The Houston/Galveston metropolitan area is a port destination for imported oil, has sizeable exports of refined products and petrochemicals, and is dominant in upstream energy exploration and development industries.

Table 4.1 identifies the current industry composition within each region and shows the extent to which the industrial composition of each region differs from the industrial composition of the nation as a whole. Sectors that will be significantly affected either positively or negatively by the expansion of hydrogen markets are noted.

Table 4.1. Industry Shares of Employment in the Regions, 2005

	Upper Midwest	Lower New England and Upper Mid-Atlantic	California	Tennessee	Houston
Over Represented	Auto/Auto Parts	Chemicals	Refining	Auto/Auto Parts	Petrochemicals
	8.0%	3.3%	2.0%	5.5%	11.0%
	Chemicals/ Plastics		Natural Gas Dist	Chemicals	Refining
	4.4%		1.0%	3.0%	9.5% Oil & Gas Extract
					9.5%
Under Represented	Oil & Gas Extract	Oil & Gas Extract	Auto/Auto Parts	Oil & Gas Extract	Auto/Auto Parts
	0.3%	0.2%	0.6%	0.1%	0.2%
	Refining	Refining	Power Generation	Power Generation	
	0.9%	0.4%	0.6%	0.2%	
		Auto/Auto Parts		Refining	
		0.5%		0.5%	

Source: Bureau of Labor Statistics, accessed August-September 2006 at http://www.bls.gov/cew/home.htm.

The Upper Midwest has a much larger proportion in automobile and automotive parts related employment than the nation as a whole. Although it has a larger concentration of chemical sector related employment than the nation as a whole, the Lower New England and Upper Mid-Atlantic region is otherwise fairly representative of the nation as a whole. Tennessee is somewhat similar to the Upper Midwest. California, with an economy larger than that of many countries, is more diversified than the other regions, although its share of employment in refining is higher than the nation as a whole. Houston is the least representative of the regional areas. The oil and gas extraction sector is nearly 10 times more important to Houston than it is to the national economy, and the regional share of automobile and automobile parts manufacturing is only a small fraction off the national average.

Table 4.2 gives the percent difference in a sector's share of regional output from the sectoral share of national employment. For example, the chemical sector's share of total employment in the Lower New England and Upper Mid-Atlantic region is 47% greater than the national share, but its oil and gas extraction sector's share of output is 89% smaller than the national share. These statistics provide the context of these regions' economic differences.

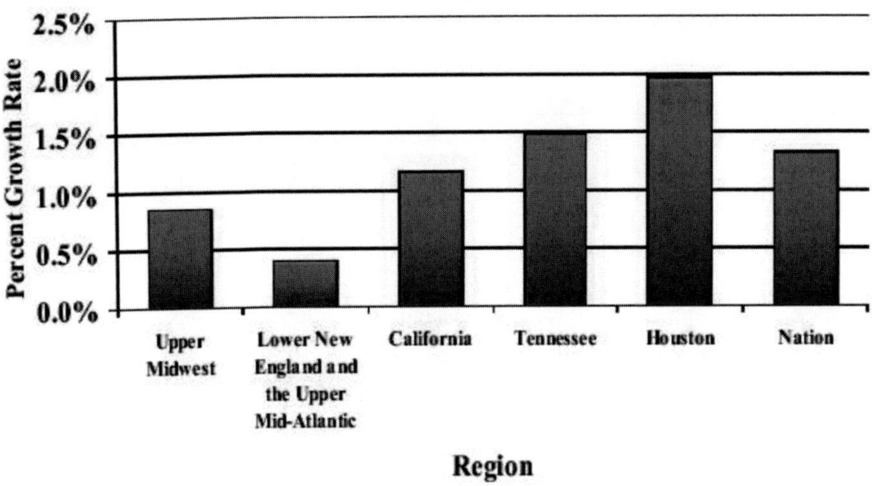

Source: Bureau of Labor Statistics, accessed August-September 2006 at http://www.bls.gov/cew/home.htm.

Figure 4.1. Average Annual Employment Growth: 1990 - 2006

Not only do the five regions analyzed here have different proportions of key sectors, but their growth rates have also differed over the past 15 years, as shown in Figure 4.1. These growth rates give a context for the hydrogen scenario impacts. If a region experiences an additional 2% employment by 2050 because of a stimulus from the hydrogen economy, that impact should be placed within the context of the overall gains the region was expected to experience without hydrogen. For a region that has been experiencing ½ % growth per year, a 2-percentage-point addition represents about 4 years of growth, which is a significant impact. For a region that has been growing at 4% per year without hydrogen, the impact of a 2-percentage-point addition because of hydrogen would constitute only 6 months of normal growth. The same need for context is true for growth within sectors.

Table 4.2. Percentage Differences between Regions' Shares of Industry Employment and National Shares, 2005

	Upper Midwest	**Lower New England and Upper Mid-Atlantic**	**California**	**Tennessee**	**Houston**
Over-Represented	Auto/Auto Parts	Chemicals	Refining	Auto/Auto Parts	Petrochemicals
	253%	47%	40%	146%	281%
	Chemicals/ Plastics		Natural Gas Dist	Chemicals	Refining
	36%		150%	36%	576%
					Oil & Gas Extract
					1089.00%
Under-Represented	Oil & Gas Extract	Oil & Gas Extract	Auto/Auto Parts	Oil & Gas Extract	Auto/Auto Parts
	-91%	-89%	-75%	-89%	-91%
	Refining	Refining	Power Generation	Power Generation	
	-34%	-68%	-53%	-81%	
		Auto/Auto Parts		Refining	
		-77%		-63%	
	For example, in the "over-represented" portion of the table, the auto/auto parts sector's share of employment in the Upper Midwest is 253% greater than the nation's share of employment in auto/auto parts; in the "under-represented" portion, the Upper Midwest's share of employment in oil and gas extraction is 91% smaller than the nation's share in oil and gas extraction.				

Source: Bureau of Labor Statistics, accessed August-September 2006 at http://www.bls.gov/cew/home.htm.

The Upper Midwest grew an average 0.85% per year between 1990 and 2006. Despite sluggishness of the region's auto sectors, manufacturing in this region fell less than the national average, and the region shared to some extent in the national growth in non-manufacturing industries. The slowest growing of the five regions was the Lower New England and Upper Mid-Atlantic region. The average growth in employment over the 16 years was 0.4% per year. The region's employment growth was impeded by a nearly 3% per year decline in manufacturing, though strong growth occurred in professional services, particularly scientific and technical services, in which all regions experienced solid growth. California grew at 1.2% per year, a rate also affected by the real estate crash in the early 1990s and the high tech crash during the first part of the present decade. Tennessee experienced solid growth during the 16-year period. Like most regions, Tennessee lost manufacturing jobs, but its losses were smaller than in other regions because of growth in auto assembly and auto parts manufacturing. Among the 5 regions studied, Houston experienced the highest average growth rate, nearly 2.0% per year. This growth occurred despite the recession of 1991-92 and the energy mini-bust in 1998-99. Houston shared in the national growth in scientific and technical services, and while non-durable manufacturing declined, durable goods manufacturing held up quite well. Upstream energy employment (exploration) has been a source of growth. Since 1999, downstream energy (refining and petrochemicals) has experienced moderate job losses.

4.2. Impacts of Hydrogen Transformation in the Five Regions Under the HFI Scenario

4.2.1. Overview

All five regions studied are estimated to experience some additional job growth as a result of the expansion of hydrogen markets. Figure 4.2 gives the difference in jobs in 2050 between the HFI scenario and the base case, as a percent of 2050 base-case jobs. Since these regions have been experiencing significant differences in annual growth rates, Figure 4.3 puts these impacts in context by showing the ratio of difference in jobs in 2050 under the HFI Scenario compared to the yearly increase in jobs that would be experienced in the region in the absence of the additional jobs due to hydrogen. This number is equivalent to the years of additional growth gained from hydrogen market expansion if the region's growth otherwise continued at its accustomed rate.

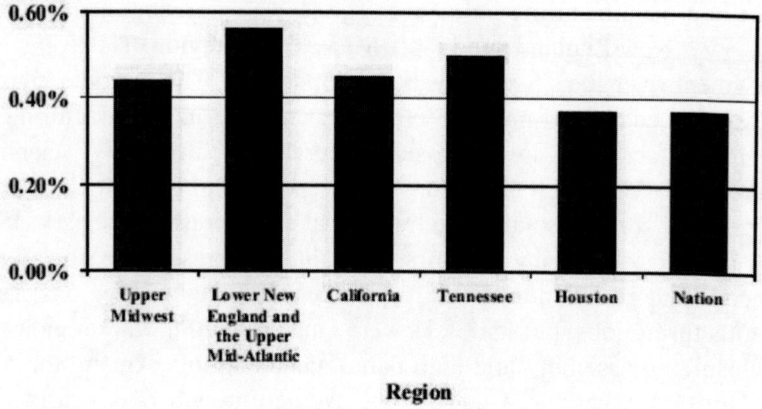

Figure 4.2. Differences in Employment between HFI Scenario and Base Case, as Percent of Base-Case Employment, 2050

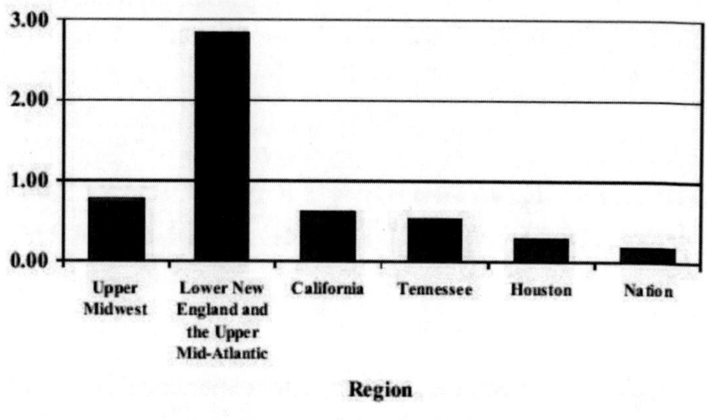

Figure 4.3. Differences in 2050 Employment between HFI Scenario and Base Case as Years of Average Annual Base-Case Employment Growth

Even under the HFI Scenario, the regional employment effects of hydrogen market expansion are modest but positive and will be spread out over more than a 30-year period. All regions will gain some additional jobs, though industries reaping the greatest employment benefits will vary across regions. Houston and California will gain the most from additional jobs in professional and technology services; the Lower New England and Upper Mid-Atlantic states will gain the most from the production and delivery of hydrogen; and Tennessee and the Upper

Midwest region will see modest gains more evenly spread across a wide array of sectors.

Table 4.3. Greatest Employment Impacts by Sector and Region

Sectors	Regions
Greatest Gainers	
Hydrogen Production	All Regions
Professional & Technical Services	All Regions
Greatest Losers	
Fabricated Metals	Upper Midwest
Downstream Energy	California
Upstream Energy	Houston
Downstream Energy	Houston

All regions likewise enjoy a modest boost in employment from the production and distribution of hydrogen under the HFI Scenario. All regions also experience noticeable impacts of the hydrogen transformation on professional and technical services, which expand to meet the technical and engineering needs of transforming the auto and refining industries and of creating a mass production hydrogen industry. Losses in energy and fabricated metals occur in some regions. Some of the major results are summarized in qualitative terms in Table 4.3.

4.2.2. The Upper Midwest Region

The fact that the Upper Midwest region is a center of automobile and automobile parts manufacturing does not mean that its employment will necessarily be significantly affected. Because the switch to hydrogen vehicles is not projected to affect overall labor productivity in the industry, the auto sectors are not expected to be significant net gainers or losers of jobs as a result of the transformation to hydrogen. In the present report, the shares of U.S. automobile production provided by U.S. companies and multinationals are assumed to be the same in the base case and in the two hydrogen scenarios. As discussed in Section 5 on international effects, differences in the relative performance of companies in introducing and marketing hydrogen vehicles, toward which the present report takes a neutral view, is an uncertainty that could affect the regional distribution since companies differ in their regional concentration of production.

The Upper Midwest, like the Lower New England and Upper Mid-Atlantic region, is an importer of gasoline, so the shift to hydrogen will promote employment gains from the transformation as the region shifts from gasoline importation from other states to hydrogen production within the state. Still, the stimulus of hydrogen production and delivery does not have a very large impact under either scenario. Although the Upper Midwest is not a major refining region, the sector experiencing the greatest hydrogen-related losses in the region is refining. A small number of other types of upstream-energy-related jobs are lost in the Upper Midwest.

In total, under the HFI Scenario, roughly 4,800 job losses occur in the region, scattered across 14 sectors, compared to 110,000 jobs created across 41 sectors, for a net creation of 105,000 jobs, or 0.44% of the region's base-case employment in 2050. Thus, compared to the nationwide employment creation of 0.37% under the HFI Scenario, the Upper Midwest region experiences modestly more expansion than the nation on average. Expressed alternatively, the net jobs created in the Upper Midwest account for 0.06 points of the nationwide 0.37%.

4.2.3. The Lower New England and Upper Mid-Atlantic Region

The Lower New England and Upper Mid-Atlantic Region experiences the largest percentage gains in employment as a result of the switch to hydrogen, with 0.56% greater employment in 2050 under the HFI Scenario than in the base case, accounting for 0.08 points of the national total of 0.37%. Given the modest growth of this region over the past decade and a half, this amounts to almost 3 years' growth.

The primary stimulus to the Lower New England and Upper Mid-Atlantic Region stems from the production and delivery of hydrogen to a large and relatively dense population. The region is not projected to experience any serious employment losses as a result of the hydrogen transformation, though some minor losses are projected in upstream energy, tied to corporate energy office workers. The minimal amount of losses helps keep the net gain high. Essentially, the gains to this region stem from the fact that it is currently a major importer of gasoline. Thus, the move to hydrogen transforms this economy from an energy importer to an energy producer, at least for its automotive needs.

4.2.4. California

California's economy also is stimulated by the new hydrogen industry. As a result, the state is projected to experience employment decreases in downstream energy, which includes the refining sector. California's high-tech sectors will participate in the development of the new technology needs of hydrogen systems. Because of the state's strength in science and engineering, California should experience some employment growth induced by a shift to hydrogen vehicles, although most production of these autos will occur elsewhere.

Overall, California is projected to see an additional 0.45% employment over the all-gasoline scenario by 2050 in the HFI Scenario, approximately equivalent to 8 months' growth under the base case. California's job creation contributes 0.04 points to the national total of 0.37%.

4.2.5. Tennessee

Tennessee's economy will be stimulated by hydrogen production, though somewhat less than other regions studied, because of a smaller population and lower population density. Tennessee's net job gain under the HFI Scenario of roughly 14,500, or 0.5% of its base-case employment in 2050, is second highest among the five regions, primarily because Tennessee has no significant losses to dampen the gains in hydrogen production and delivery, and technological and engineering services. The state's employment growth accounts for 0.01 points of the 0.37% nationwide employment creation.

4.2.6. Houston

Houston gains employment at the same rate of the nation as a whole. However, given its large percentage of employment in the petrochemicals, refining, and oil and gas extraction sectors (see Table 4.2). Houston is the region that has the largest share of employment at risk. Because of this fact and because Houston is only a single metropolitan area with a smaller population than the other areas studied, Houston gains the least of any of the five regions by 2050 in absolute numbers of new employees. Thus, Houston's gains contribute only 0.005 point to the total national employment gain of 0.37% under the HFI Scenario. This is representative of oil dependent economies, even though Houston's energy base is quite diverse. Exploration activity will, on net, be hurt by the shift to hydrogen systems, though much of Houston-based exploration is tied to the search for natural gas, which will be modestly stimulated. Houston's refining will be deleteriously affected, but this sector has not been a source of growth since the late 1990s and is not expected to be so in the future under an all-gasoline scenario. The key for this sector will be the extent to which it transforms its product mix away from gasoline and toward other products.

On the positive side, Houston has been producing hydrogen for decades for use in refining and petrochemical processes. Though roughly half of the production of hydrogen is assumed to come from coal in the hydrogen adoption scenarios, the technical expertise in Houston in energy, chemicals, and gases has the potential for applications in the hydrogen economy. Houston firms are already involved in a wide array of non-oil-related energy projects and activities. Thirty percent of Houston's increase in employment under the HFI Scenario, for example, will be in the high-tech-oriented professional and technical services sector.

During the 1990s, the non-energy portion of Houston's economic base grew more than three times as fast as its combined upstream and downstream energy, thereby reducing energy's share of the region's economic base below 50%. Hydrogen market expansion will likely help this trend continue.

Chapter 5

INTERNATIONAL COMPETITION

Two major objectives will be served in the international arena. First, reduction in oil imports, with the attendant increase in energy independence, is a clear national goal to which the hydrogen economy will contribute. Second, if U.S. companies are able to forge a lead in hydrogen technologies, U.S. global competitiveness will be fostered. In light of these objectives, this section considers effects on U.S. production of hydrogen-related products and hydrogen infrastructure, energy imports, multinational corporations, the balance of payments, and capital markets.

5.1. U.S. SHARE OF HYDROGEN-RELATED PRODUCTS

5.1.1. Vehicles and Vehicle Parts

The high cost of shipping automobiles and automobile bodies dictates that body manufacture and assembly of the vast majority of vehicles will continue to be located in the countries of final demand. The location of vehicle parts manufacturing is much less tied to final demand location.

Present patterns of imports and exports in vehicle parts provide a basis for estimating how U.S. competitiveness in parts manufacture would be affected by a transformation to hydrogen vehicles. Table 5.1 reports the value of trade in vehicle components, by component group, in 2004. Engine-related components (including complete engines, air and fluid handling, and electrical parts), the largest traded component group, accounted for \$21.3 billion of the \$68 billion in imports in 2004. Trade in transmission and powertrain components and storage batteries was also dominated by imports, accounting for 50% of U.S. vehicle component imports and 35% of exports.

Table 5.1. Value of U.S. Trade in Vehicle Components, by Component System, 2004 (in 2005 Prices)

	Imports			Exports		
System	**Value, $ billion**	**% of Total Imports**	**% of Domestic Production**	**Value, $ billion**	**% of Total Exports**	**% of Domestic Production**
Engine	21.338	31%	32%	9.015	21%	14%
Gasoline Engine and Engine Parts	12.651	19%	31%	5.871	13%	14%
Electronic Equipment	8.687	13%	34%	3.145	7%	12%
Transmission and Powertrain Parts	11.090	16%	24%	5.388	12%	12%
Storage Battery	1.948	3%	46%	0.766	2%	18%
Body	1.135	2%	3%	1.817	4%	5%
Motor Vehicle Metal Stampings	0.518	1%	2%	1.643	4%	6%
Motor Vehicle Bodies	0.617	1%	5%	0.174	<1%	1%
Steering and Suspension	3.845	6%	26%	1.322	3%	9%
Brake System	3.771	6%	21%	2.198	5%	12%
Interior	5.474	8%	19%	2.343	5%	8%
Seating and Interior Trim	3.873	6%	18%	1.780	4%	8%
Air Conditioning	1.601	2%	21%	0.563	1%	7%
All Other Parts Manufacturing	19.760	29%	43%	21.036	48%	46%
Total	68.361	100%	26%	43.886	100%	17%

Source: U.S. International Trade Commission: accessed August-September 2006 at http://dataweb.usitc.gov/.

Canada, Japan and Mexico were the countries of origin for 72% of all auto parts imported into the U.S. in 2004, with particular concentrations in engine components. Japan has been the leading exporter of powertrain parts, and

Japanese domestic producers remain closely tied to Japanese multinational vehicle producers located in the U.S. Most U.S. exports of engine and powertrain parts are destined for Canadian assembly plants of DaimlerChrysler, Ford, and General Motors. The recent decline in these U.S. parts exports to the Canadian plants largely reflects the loss in U.S. motor vehicle market share.[1]

The components in the fuel cell system can be categorized into the fuel cell stack, the fuel cell balance-of-plant (consisting of heat exchangers, controller, and air compressor or blower), and the hydrogen storage tank. Manufacture of the fuel cell stack involves new materials combined with traditional manufacturing processes tailored to fuel cells. Development of these processes is being pursued in the U.S., Japan, Korea, and the U.K. primarily, although China also has development programs. Once developed, the manufacturing processes could be moved readily to low-cost countries. Pressure to do so could be lessened because the largest cost contributor, the electrodes, will be made in automated processes, which would reduce their labor cost. Developers may want to keep manufacturing in-house to protect the intellectual property and proprietary knowledge. The high contribution of materials to the cost would tend to reduce the drive to move off-shore. The balance-of-plant components can easily be made in low-labor-cost countries, although the size of heat exchangers might favor their being made locally. Controllers and compressors can be made anywhere. At this time, hydrogen storage technology is an active area of research with compressed hydrogen available for purchase. Composites are made worldwide, and this should not be different for hydrogen storage tanks. Safety requirements and manufacturing controls may be considerations that keep manufacturing domestic for some period of time. The United States, Canada, and Japan each currently have a domestic leader in tank development. While any technology can be made in low-labor-cost manufacturing regions, maintaining control of intellectual property and continued development of the processes would tend to keep manufacturing of an array of components in the U.S.

Companies of many nations are competing in the emerging technologies that will be used to produce materials and components for hydrogen vehicles. Many of these companies, American as well as foreign, have manufacturing facilities in their home countries as well as abroad, and they will supply foreign markets through combinations of exports from the home country and production in the host countries. There may be some tendencies to supply more exports in early years of hydrogen market expansion (around 2020) to keep closer control of intellectual property, but even this tendency may be diluted by multinationals' location of R&D facilities in the U.S. The scope for sending maturing, labor-intensive production processes and products to lower-income countries is not expected to be substantially different for hydrogen vehicle components than for conventional components, and that could characterize issues of concern around 2035. Under either the HFI or Less Aggressive Scenario, by 2050, the vehicle elements of the hydrogen market will have matured. The effects of hydrogen market expansion on the location of automobile parts production depends on considerations such as the likely prospect of continuing change in the technologies, the experience of the companies in the automotive industry with evolving platforms and component systems, and specific developments in the many components and technologies over which competition will take place. While individual parts could be affected favorably or unfavorably, there are no indications that U.S. exports or imports of parts as a whole will be significantly affected by hydrogen market expansion.

However, a hydrogen transformation may significantly affect U.S. competitiveness in specific auto parts. The hydrogen fuel cell system will completely replace the engine, transmission, and powertrain parts and will modify storage batteries. Foreign producers have dominated these conventional, gasoline-vehicle components in recent years. Substitution of hydrogen fuel cell technology for gasoline systems could give U.S. manufacturers of the hydrogen systems an opportunity to re-capture recently lost market shares in vehicle components. With the exception of platinum, most of the materials used to manufacture fuel cell systems can be sourced domestically. Current and likely future sources of materials and components are as follows:

Carbon fiber (gas diffusion layer and hydrogen storage tank). Two grades of fiber are used in the overall system. Within the stack, the carbon fiber used in gas diffusion layers must be of high purity and be graphitized for stability. There would be many options to source this type of fiber domestically and internationally. The carbon fiber used in compressed hydrogen storage tanks must be aerospace grade with a high strength and quality to ensure reliability. While several sources of aerospace grade material exist, Toray of Japan is currently one of the dominant suppliers. In this industry, when large quantities of material are needed for a dedicated customer (e.g., by AirBus), suppliers (e.g., Toray) have built local plants to meet these needs. With the high volumes anticipated for transportation fuel cell vehicles in the U.S., suppliers would probably do the same in a hydrogen economy.

Fuel cell membrane materials (the electrolyte). Several options are available for manufacturers of electrolyte membranes in the U.S. DuPont is currently the largest supplier. 3M and Gore are also major companies producing membranes.

While there are several companies in Japan and Europc supplying membranes, U.S. manufacturers are very competitive in this area.

Graphite powder (for bipolar plates). Within the stack, the bipolar plates that electrically connect the individual cells within the stack are the largest mass of material. Graphite powder can be sourced both domestically and overseas. The resins that bind the graphite powder together can also be sourced in the U.S. Additionally, developers are considering using metal bipolar plates as an alternative to graphite bipolar plates. These metals and the required surface treatments can be obtained domestically.

Electrode materials (primarily platinum). Platinum is the dominant material in the electrode and would have to be imported. Catalyst supports used in the fuel cell may be manufactured in the U.S. or by foreign sources. South Africa is the dominant supplier (approximately 80%). As the number of fuel cell vehicles increase, recycled platinum will become an important source of material, and the volume of imports will decrease significantly. The recycling plants probably would be sited in the U.S. for proximity to the recycled materials.

Other materials for stack hardware. All other materials, such as metal or composite endplates, bolts, and wires, and balance-of-plant components (e.g., blowers, heat exchangers) could also be obtained domestically. The raw materials for these components could be produced domestically or imported.

Compressed (on-board) hydrogen storage. Several companies in North America have developed carbon fiber composite high pressure tanks for hydrogen storage. Toyota has also made announcements concerning their own technology developments. Due to the large number and size of compressed hydrogen storage systems, it is anticipated that tanks and systems would be manufactured in proximity to the automotive plants.

High-tech batteries. Japan, China, and Taiwan now dominate the manufacturing of high-tech batteries for digital electronics. A partnership of Panasonic and Toyota make the Prius nickel metal hydride battery today. Adoption of hybrid electric vehicles or plug-in hybrid electric vehicles would significantly stimulate the production of advanced batteries in these countries.

5.2. Hydrogen Infrastructure

Different opportunities will arise in the manufacture of infrastructure components and construction and operation of hydrogen production and delivery facilities. Construction and operation will be conducted domestically, but manufacture of components will involve a combination of imports and export opportunities.

5.2.1. Infrastructure Construction

Because of the difficulties in transporting it, hydrogen will not be an internationally traded commodity, with the possible exception of some relatively short-distance, cross-border trade with Canada and Mexico. Consequently, the production facilities serving the U.S. market will be built and operated domestically. Foreign participation in the construction of the large, centralized hydrogen generation facilities is likely to resemble current involvement in U.S. refinery or chemical plant construction. The bulk of the engineering contractual work would likely be undertaken by domestic companies. While some foreign firms could design some of the generating facilities, they would have to contract out the actual construction work to U.S. companies. Most of the cement would be provided by local firms, while most of the steel probably would be imported. The U.S. currently has a strong position in supplying reforming equipment for domestic operations, and the U.S. shares the provision of domestically operated electrolyzers with Canadian and European suppliers.

5.2.2. Infrastructure Components

Because of the increase in the volume of hydrogen, the quantities of these infrastructure components will increase. The present distribution among countries of sources for hydrogen infrastructure materials and components gives a basis for assessing the extent to which hydrogen transformation will lead to relative changes in the sources.

Large container vessels. These vessels, used in centralized production facilities, use high pressures and temperatures and would require specialized alloy materials and manufacturing techniques that no longer exist in the U.S. They would have to be imported.

Pipelines. The U.S. has several large manufacturers of pipeline (Air Products, Praxair) but there is no reason to discount European and Japanese involvement in hydrogen pipeline manufacturing when pipeline demand reaches large scale. Given the large capital outlays, any hydrogen pipeline manufacturer would probably be looking for partners, and those are likely to be domestic or foreign oil or gas companies.

The smaller pipelines are likely to be produced in the U.S., while many of the larger pipelines could be manufactured in Canada, which possesses a current advantage given its large natural gas industry.

Carbon fiber and aluminum liner. Unlike natural gas, hydrogen causes embrittlement and metal fatigue. Light-weight materials like aluminum liner will be used to contain the hydrogen inside any storage facility, pipeline or cylinder. Carbon fiber will be used to add strength to the liner. The carbon fiber, even if produced by a foreign company, probably would be located near its mature markets, and aluminum liner could be supplied domestically.

Dispensers (at fueling stations). At least one-half of dispensers are currently manufactured in the U.S. Most of the metering equipment, however, is imported.

Compressors. Large compressor stations for hydrogen are currently manufactured both domestically and abroad. Europe, Japan and Canada are expected to remain competitive.

Turbines and generators. These large capital goods contain numerous components, many of which are sourced from both the U.S. and abroad. Germany, Japan, the U.S. and eventually China are expected to be competitive in this product field.

Valves, bolts, threads, and well fittings. These products are made abroad, in both Europe and Asia. The exception is large valves, which are manufactured in the U.S. This pattern of comparative advantage is not foreseen to change.

5.3. Energy Imports

5.3.1. Oil

In 2005, oil imports accounted for 1.5% of gross domestic product, nearly 11% of all imports, and 24% of the trade deficit.[2] Light-duty vehicle oil consumption in 2020 is estimated at $300 billion, with two-thirds or $200 billion being imported. Assuming further increases in oil consumption will have to come primarily from increased imports, growth in demand due to growth in population, vehicles per household and miles driven per vehicle would add $79 billion to the import bill by 2050.[3] Adding the $79 billion to the $200 billion gives a total 2050 import savings of $279 billion at 2020 prices if gasoline-powered light-duty vehicles were eliminated. The world oil price is projected to rise 38% between 2020 and 2050, so at 2050 prices, the import savings would be $385 billion. Under the HFI Scenario, ninety-six percent of the light-duty vehicle stock is expected to be hydrogen in 2050. Ninety-six percent of the $385 billion is $370 billion in import savings from going to hydrogen. This estimate assumes no increase in fuel efficiency in gasoline vehicles, which to the extent it occurs will reduce import savings. Import savings would be lower in the years before 2050 and also lower under the Less Aggressive Scenario—$229 billion in 2050.

5.3.2. Natural Gas

The volume of natural gas imports nearly tripled between 1990 and 2005, and their value increased by a factor of 7 because of the price increases over that period. Imports of natural gas accounted for 1.7% of all U.S. imports in 2005.[4] With the assumption that coal gasification with carbon sequestration and biomass will be used in preference to reforming of natural gas, the demand for natural gas will not see an appreciable increase due to large expansion of hydrogen markets. Some natural gas will be used during early market growth and may continue to be used in some cases, for stationary fuel cells, for example; however that would be offset to some extent by diversion of natural gas demand from electric utilities as fuel cells supply a portion of incremental electricity demand.

5.4. MULTINATIONAL CORPORATIONS

Continued U.S. federal investment in hydrogen R&D will foster global competitiveness of U.S. firms as the economy is transformed to hydrogen. The movement to hydrogen could well be an opportunity for U.S. automotive firms to recapture market share lost to foreign multinationals in recent years.

The backdrop is the on-going growth in production of U.S. corporations in other countries and of foreign multinationals in the U.S., which is a part of economic globalization. These changes affect where assets are owned, but they do not necessarily have a great effect on the location of production and employment, which is determined largely by underlying cost and demand considerations. U.S. firms have successfully established vehicle assembly plants in India and China, which will be the world's major purchasers of light-duty vehicles by 2030, while Japanese, Korean, and German firms have located plants in the U.S. Competition among U.S. companies and multinational producers in other countries can be expected to continue with hydrogen vehicles. R&D in hydrogen vehicle technologies is occurring worldwide, and it can be expected that most major vehicle and components manufacturers will find the new technologies available to them. As research becomes more applied, companies may follow different development paths, though any predictions about the relative success of companies of different national origins would have a weak basis.

Two areas where effects could be noticeable, however, stem from the choices of multinationals as to where to carry out R&D—which could in turn affect the location of parts production—and differences between the choices made by of U.S. corporations and multinationals regarding the location of their automobile production facilities.

Foreign vehicle multinationals have begun conducting significant automotive research and development in the U.S., for two principal reasons: to take advantage of America's relative abundance of engineering talent and its engineering education facilities, and to ensure their vehicles meet U.S. environmental and safety standards. Japanese firms have led the way, with 34 U.S-based R&D centers employing more than 3,000 in 2004, up from 200 in 1987.[5] These U.S.-based facilities are being given more responsibility over time. For example, the 2005 Avalon sedan is the first Toyota vehicle engineered from start to finish at its Toyota's Ann Arbor facility. Foreign firms can be expected to locate R&D facilities for hydrogen vehicle technology in the U.S. for the same reasons their R&D centers are growing rapidly today. This development could offset the tendency among foreign firms to keep new technology developments close to home, with concomitantly greater production of both automobiles and parts in the U.S.

5.4.1. Future Hydrogen Production by Multinationals

Both foreign and American multinationals are well represented in the energy and industrial gas manufacturing industries. The multinational oil companies are likely prospects to produce hydrogen, particularly from centralized facilities. Their hydrogen production activities overseas probably would parallel those in the U.S.

The industrial gas manufacturing firms also include both U.S. and foreign multinationals: the American Air Products and Praxair in the United States, Linde in Germany, and Air Liquide in France. Linde recently purchased British BOC Gases to become the world's largest industrial gas company. Each of these firms will acquire shares of the expanding U.S. hydrogen market, but there is little basis for predicting changes among their shares of the market.

The construction of the reactors to provide nuclear-generated hydrogen likewise could involve both American and foreign multinational reactor vendors. The operation of the facilities would be strictly domestic, however.

5.4.2. Regional Implications

The location of foreign multinationals' plants in the U.S. has changed the distribution of the vehicle industry from the Upper Midwest and along the east coast toward the South, in a north-south strip largely between Interstates 65 and 75, with allowance for new plants in South Carolina.[6]

The past twenty years' experience of foreign multinationals' locational choices in the U.S. is a reasonable guide to their participation in hydrogen vehicle components and assembly. The future division of production in the U.S. between domestic and foreign multinational corporations may continue to affect the regional distribution of production. If so, the extent to which U.S. corporations or foreign multinationals take the lead in introduction of hydrogen vehicles could affect the location of automobile production within the U.S., with implications discussed above in Section 4 on regional impacts.

5.5. Indirect International Effects on the U.S. Economy

5.5.1. Effects on the Trade Balance and Exports

The \$370 billion oil import savings estimated above under the HFI Scenario in 2050 amounts to 1% of the \$38.12 trillion projected 2050 gross domestic product. Some of the reduction in expenditures on imported oil will be redirected to a domestically produced, largely nontradable good—hydrogen. The \$370 billion reduction in oil imports equals approximately 7.5% of projected total U.S. imports in 2050 if trade grows at the same rate as gross domestic product. The oil import reduction introduces a trade imbalance that will lead to a combination of adjustments in non-oil imports and exports. The adjustments will be spread over the many non-oil commodities involved in U.S. trade. Considering that both import and export adjustments will occur and that they will occur over a 30-year hydrogen market expansion period, the average yearly quantity adjustment for a typical commodity will be less than 0.1 of one percent. Similar considerations apply to all countries adopting hydrogen. On-going globalization effects on trade over the coming years seem bound to overwhelm the small trade adjustments to the reduction in oil imports.

5.5.2. Effects on the World Capital Market and Interest Rates

A similar conclusion applies to effects on world capital markets. Some of the oil producing, and particularly OPEC, countries have had high propensities to save their oil profits and have invested extensively in industrialized nations' assets. In 2005, OPEC dollars recycled into net direct and portfolio investments, worldwide, amounted to around \$100 billion.[7] With falling OPEC incomes, OPEC countries' purchases of U.S. assets would be reduced.

Overall financial lending to industrialized countries would be lowered. Because OPEC demand for financial assets is a small part of total world demand for assets, any effect on interest rates might not be perceptible. The fact that the effect is spread over many years would make it even less noticeable, particularly since non-hydrogen events can be expected to have many and varied influences of much greater magnitude on world interest rates.

End Notes

[1] Klier, T.H., Rubenstein, J.M., "Competition and Trade in the U.S. Auto Parts Sector," *Chicago Fed Letter,* No. 222 (January 2006).

[2] Trade data: Office of Trade and Industry Information (OTII), Manufacturing and Services, International Trade Administration, U.S. Department of Commerce, accessed August – September 2006 at http://ese.export.gov/; oil imports data: Energy Information Administration, accessed August – September 2006 at http://tonto.eia.doe.gov/dnav/pet/pet_move_impcus_a2_nus_ep00_im0_mbbl_m.htm; balance of payments data: Bureau of Economic Analysis, accessed August – September 2006 at http://www.bea.gov/bea/di/table1.xls; GDP data: http://www.bea.gov/bea/dn/home/gdp.htm.

[3] $79 billion is derived from 0.78% growth award factor to the total annual.

[4] Energy Information Administration sources: accessed August – September 2006 at http://tonto.eia.doe.gov/dnav/ng/ng_move_impc_sl_a.htm and http://tonto.eia.doe.gov/dnav/ng/ng_move_expc_sl_a.htm.

[5] Japan Automotive Manufacturers Association (JAMA), "Growing in America," accessed August – September 2006 at http://www.jama.org/library/factsheets/Contributions_2005.pdf, pp.1, 4.

[6] Klier, T.H., Rubenstein, J.M., "The Supplier Industry in Transition—The New Geography of Auto Production," *Chicago Fed Letter,* No. 229b (August 2006).

[7] International Monetary Fund, *World Economic Outlook – Globalization and Inflation,* Washington D.C., April 2006, Figure 2.3, p. 74.

Chapter 6

CONCLUSION

This report estimated the employment impacts of a transformation of the U.S. economy to hydrogen economy between 2020 and 2050. The report examined the differences in employment by industry between a non-hydrogen base case scenario and two scenarios of expanded market penetration of mobile and stationary hydrogen fuel use. The more rapid transformation scenario followed the HFI of saving 11 million barrels of oil per day by 2040, and the less rapid scenario followed DOE's analysis supporting its 2006 program benefits estimation. Under the HFI Scenario, the penetration of the light-duty vehicle stock with hydrogen powered vehicles by 2050 is 96%. Under the Less Aggressive scenario, penetration by 2050 is 38%. The choices of hydrogen production technologies supplying the demands for hydrogen at the three dates examined—2020, 2035, and 2050—and the shares of production contributed by each, were based on DOE's H2A models.

National employment impacts were estimated from the 509-sector IMPLAN inter-industry model for the U.S. economy. Revised industry purchase vectors were constructed for three industries for the hydrogen scenarios—vehicle components, vehicle assembly, and hydrogen production. By 2050, under the HFI Scenario, the transformational adjustments are fully completed, with no more anticipatory investment; U.S. employment is increased by a net of 0.37%, or 675,000 jobs out of a total projected base-case employment of 184 million. Under the Less Aggressive scenario, the transformation is not fully completed by 2050; U.S. employment is increased by that year by a net of 0.20%, or 361,000.

Projections of the job creation and job replacement underlying the total employment changes were based on expected changes in the occupational and skill structure of directly affected industries, estimated with the assistance of industry opinions. While net employment in the automotive industry is unchanged between the gasoline and hydrogen economies, replacement of gasoline-related skills with hydrogen-related skills is substantial under the HFI Scenario. In automotive parts manufacturing, 12,000 white collar jobs are created by 2050; and blue-collar job creation and replacement is 117,000 by 2050. The greatest job creation and replacement is for automobile dealerships and repair: 680,000 by 2050.

Training implications of the job changes vary by industry and skill, as indicated by a survey of industry opinions. Most of the needs for new skills can be supplied by normal rates of entry into the labor force as workers receive training in new, hydrogen-related skills. Considering the small proportion of all engineers in the U.S. that the automobile industry employs, engineering schools have the capacity to respond to changes in the fields of engineering. Up to 110,000 technicians and mechanics will need to be equipped with hydrogen-technology skills by 2020 to service new vehicles in the HFI Scenario. By 2050, the number is 630,000, many of whom will have entered the labor force prior to the beginning the hydrogen transformation.

Regional employment impacts were estimated using regional differences in industry structures and resource bases for hydrogen production. Compared to the base case without a hydrogen transformation, the HFI Scenario leads to a projected gain in the Upper Midwest of 105,000 net additional jobs by 2050, or 0.44% of its base-case employment, distributed across 41 industries. Projected increases in employment over the base case are 0.56% for the Lower New England and Upper Mid-Atlantic region, 0.45% for California, and 0.5% for Tennessee. Houston's energy experience leads to employment gains in hydrogen production and pipeline equipment, with an overall 0.37% gain over the base case.

The employment impacts of a hydrogen transformation on international competitiveness are limited by the fact that most vehicle production will continue to be tied to countries with large automobile demands. If the shares of U.S. production provided by U.S. multinational corporations are affected because some companies get ahead of others in introducing hydrogen vehicles, the location of automobile production within the U.S. could be affected due to the fact that companies differ in their regional concentrations of production facilities.

Recommendations

This study identified possible employment impacts that could result from hydrogen market expansion in the transportation, and stationary and portable power sectors. As noted previously, any study of potential future impacts necessarily presents difficult challenges and involves significant uncertainties. Results and recommendations should be considered with those issues in mind.

The scenarios, hydrogen generation options, and regions selected for the study yielded a reasonable measure of the potential opportunities that hydrogen presents to U.S. employment. The study considers introduction of the fuel cell vehicle and supporting hydrogen infrastructure development that spans a period of approximately 40-50 years. The study highlights possible skill and education needs to support the associated industries and technologies. In addition to the specific skill requirements of the fuel cell industry, future education of the next generation should be focused on skill sets to have the ability to adapt to changing technologies.

Training implications of the job changes vary by industry and skill, as indicated by a survey of industry opinions. Most of the needs for new skills can be supplied by normal rates of new entry into the labor force as workers receive training in new, hydrogen-related skills. Considering the small proportion of all engineers in the U.S. that the automobile industry employs, engineering schools have the capacity to respond to changes in the fields of engineering. Following are employment-related recommendations of a transition to a hydrogen economy.

1. Training programs

a. Training and retraining programs may be needed to help ensure that the U.S. workforce possesses the appropriate skills and that sufficient numbers of trained personnel are available to meet the manufacturing requirements at the time that hydrogen fuel cell vehicles begin to come off the assembly lines. Development of these programs should involve close coordination between the fuel providers and auto manufacturers, and schools. At the appropriate time university and vocational programs need to be assessed to understand where the opportunities lie and what additional curricula may be needed.

b. Training and retraining programs may be useful in related aftermarket areas such as repair and recycling.

c. Educational programs aimed at the general public could help to influence people to pursue jobs in hydrogen and fuel cells.

2. Additional analysis

a. *Analysis of training needs*: The study found that training for new skills may be needed across a wide spectrum of industries. Most of the needs for new skills can be supplied by normal rates of new entry into the labor force as workers receive training in new, hydrogen-related skills. Some changes in skills appear to be relatively well defined, but many likely changes remain difficult to forecast, since many of the technologies are still maturing. Many job tasks remain unknown at present, making identification of training needs an interactive task with job definition.

b. *Assessment of skill changes, with attention to industry adjustments in different regions*: Unemployment resulting from the elimination of jobs associated with obsolete technologies and industries is estimated to be slight. The supply of labor with technology-specific skill sets may or may not keep pace with the labor demands associated with the rapid growth in the hydrogen-related industries. Even though this study indicated every region would gain jobs, surpluses and shortages of skilled workers could vary regionally.

APPENDICES

APPENDIX 1. DETAILS OF THE HYDROGEN ECONOMY SCENARIOS

Section 2 of the preceding study presented an overview of a base case and two hydrogen economy scenarios. This appendix (Appendix 1) describes (a) the model that was used to extend the base case from the years 2030 to 2050 and to estimate hydrogen vehicle sales and fuel demand, (b) the technology assumptions used to estimate hydrogen production and delivery infrastructure requirements, (c) the feedstock issues associated with alternative hydrogen production technologies, and (d) the assumptions regarding stationary fuel cells.

A.1.1. VISION MODEL

The VISION model was used to extend the base case from the years 2030 to 2050 and to estimate hydrogen vehicle sales and fuel demand. The VISION model provides estimates of energy use, oil use, and carbon emission impacts through 2050 for advanced light- and heavy-duty vehicle technologies and alternative fuels. Developed over the past decade by staff of Argonne National Laboratory for DOE's Office of Planning, Budgeting and Analysis, the model has been used for a variety of quick-turnaround requests as well as for longer-term analyses. Examples of the latter include the Hydrogen Posture Plan, the President's Hydrogen Fuel Initiative, and the 30x30 Study.[1,2,3] The VISION model consists of two Excel workbooks—a base case of U.S. highway fuel use and carbon emissions to 2050 and a copy of the base case that can be modified to reflect alternative assumptions about the market penetration of advanced vehicles and alternative fuels. The model incorporates a set of vehicle-survival and age-dependent-usage procedures to track vintage-specific vehicle stock and usage. The model then estimates light- and heavy-duty vehicle stock composition, vehicle miles traveled, and energy use. The model is calibrated annually to EIA's projections in its Annual Energy Outlook, which currently extends to 2030.

For this study, VISION was recalibrated to the high oil price case of the 2006 Annual Energy Outlook (AEO). Some parameters were extended to 2050 using standard references; others were extrapolated based on trends in the Annual Energy Outlook. Energy prices were projected to reflect the relationships and trends in the AEO forecast using DOE internal models. The resulting VISION run generated the base case forecasts of light-duty-vehicle sales, total vehicle stock, and gasoline use. These forecasts provided key inputs to the IMPLAN input-output model, discussed in the preceding report and in Appendix 4 *infra*. This revised VISION base case was then run with the two hydrogen market penetration scenarios to provide alternative forecasts of light-duty vehicle sales, stock, and fuel use.

As discussed in Section 2 of the report, the two scenarios of a hydrogen economy that were examined in this study differ primarily in their assumed penetration of hydrogen fuel cell vehicles in the light-duty-vehicle market. The President's Hydrogen Fuel Initiative (HFI) Scenario uses aggressive market penetration assumptions, while the Less Aggressive Scenario uses more modest assumptions from the FY 2007 Government Performance Reporting and Results Act (GPRA) analysis.[4,5], Figure A.1.1 shows the resulting forecasts of hydrogen and gasoline demand by light-duty vehicles in the Base Case, the HFI Scenario and the Less Aggressive Scenario. In the VISION model runs, not only does hydrogen displace gasoline in both hydrogen scenarios, but total light-duty-vehicle fuel use drops by nearly 70 % (from 216 billion gge [gallons of gasoline equivalent] to 66 billion gge) in the HFI Scenario and 48% (to 112 billion gge) in the Less Aggressive Scenario.

Table A.1.1. Hydrogen Production Technologies Included in this Study

Technology	Description	Requirements and potential constraints*	Rated capacity T/d (CF%[a])	Year commercially available
Central steam methane refo-rming, with sequestration	Natural gas feedstock (450,000 lb/hr @750° F and 450 psia) is de-sulfurized, mixed with process steam and undergoes strongly endothermic reactions over Ni-based catalyst at 1,400-1,700°F. Process gas (CO & H_2) passes through a heat recovery step and a water gas shift reactor to produce additional H_2.	Industrial **natural gas** and pipeline, water	380 (90)	2015
Distributed steam meth-ane reforming	Hydrodesulfurized natural gas is steam reformed at 10-atm with PSA gas cleanup. Factory built, skid-mounted unit is assumed to achieve 75% H_2 recov-ery. H_2 exits PSA at 300 psi and is compressed and stored at 6,250 psi for cascade filling.	Industrial **natural gas** and distribution, grid electricity, water	1.5 (70)	2015
Distributed electrolysis	H_2 from high pressure (300 psi) alkaline electrolysis is compressed to 6250 psi. Process excludes O_2 capture & co-product sale.	Demineralized water, electricity from grid	1.5 (70)	2015
Central coal gasification, with seques-tration	Coal is fed to an advanced transport gasifier. Produ-ct gas is sent to a hot gas desulfurization unit (where byproduct S is produced), and a membranebased H2 separation and shift conversion unit. An advanced turbine system and solid oxide fuel cell are used to produce 171 MW of byproduct electricity. Gasifier O_2 is produced from an ion transport membrane un-it. CO_2 is compressed to 2200 psi for sequestration.	Utility steam coal and transport, water	246 (90)	2025

Table A.1.1. (Continued)

Technology	Description	Requirements and potential constraints*	Rated capacity T/d (CF%[a])	Year commercially available
Mid-size biomass gasification, no sequestration	Woody biomass (represented as hybrid poplar) is fed to a gasifier (indirectly heated with hot sand), conventional catalytic steam reformer, high- and low-temperature water gas shift reactors, and PSA. Steam is the fluidizing gas; no O_2 (or air) is fed to the gasifier.	**Cropland for woody biomass, transport, water**	164 (90)	2015
Mid-size wind electrolysis	A 276 MW wind farm with 185 x 1.5 MW wind turbines and 119 electrolyzers produces an average of 50,000 kg/d H_2. The installation sells about 324 X 106 kWh of byproduct power back to the grid annually. No O_2 is assumed to be sold.	**Land, process water, substation**	86 (58)	2015
Central nuclear thermo-chemical water splitting	Advanced high-temperature gas-cooled reactors (HTGR) (4 x 600MWt modules) provide direct process heat for the sulfur-iodine thermo-chemical process for splitting of water.	Demineralized water, uranium (if no breeder)	768 (90)	2025
Central nuclear electrolysis	Nuclear power from HTGR is used to generate electricity, some of which is used to split water and some is sold to the grid. Co-product O_2 can be sold.	Demineralized water, uranium (if no breeder), substation	719 (90)	2025

[a]CF is capacity factor or the ratio of average output to full-capacity output, expressed as a percent.

Sources: Technologies, capacity, availability: H2A Hydrogen Production Models. Capacity is in metric tons/day. Resource constraints: estimated in this study. For further discussion, see Section A.1.4. Resources in bold font could experience significant constraints.

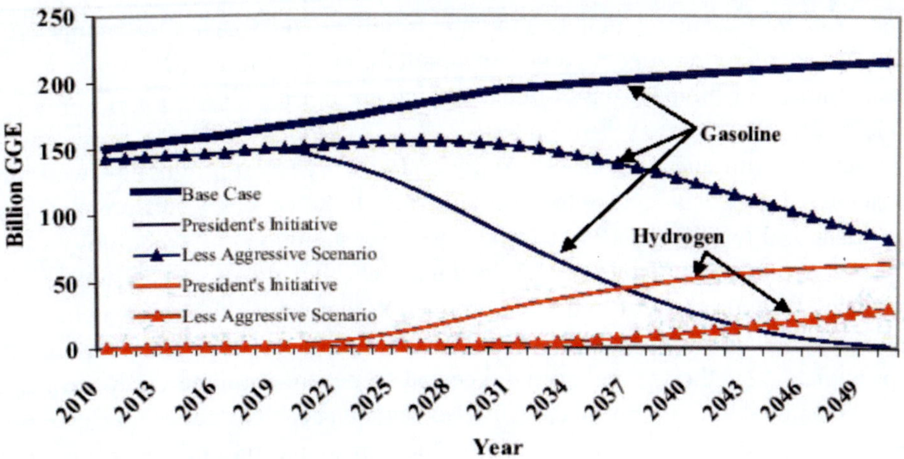

Source: Calculated by VISION model.

Figure A.1.1. Hydrogen and Gasoline Use by Light-Duty Vehicles in the Base Case, HFI, and Less Aggressive Scenarios

A.1.2. HYDROGEN PRODUCTION TECHNOLOGIES

For hydrogen production, a variety of technology options have been considered. These range from current processes that produce hydrogen by steam reforming of natural gas to processes that are still in the research and development stage, such as thermochemical water splitting using nuclear energy. The most promising potential feedstocks for hydrogen production include natural gas, coal, biomass, and water. For the first three feedstocks, energy embodied in the feedstock can be used to fuel the endothermic process of extracting hydrogen from the molecules. In the case of water, however, an additional source of energy is required. The most likely addtional sources of energy are nuclear power and wind.

Within each technology type, such as production of hydrogen from coal, many assumptions must be made in order to characterize the technology and estimate costs. These include the initial choice of technology type, as well as the specific process (e.g., gasification), its design, and the size of facility to be characterized. Some technologies, such as steam methane reforming (SMR), can operate at both small scale (forecourt[6] or distributed production) and large scale (central plant). Another technical judgment is whether to include sequestration for those options that produce a byproduct stream of carbon dioxide (CO_2). For this analysis it was assumed that central plants relying on fossil fuel inputs could sequester the carbon economically, but smaller plants could not. No sequestration was assumed for biomass-based plants, which are not net CO_2-emitters on a life-cycle basis.

Designs with and without electricity and/or oxygen co-production were also considered for nuclear and wind technologies. In some cases, different designs were analyzed for different time periods, on the assumption that technology would improve over time. Table A.1.1 summarizes the most representative and promising technologies for hydrogen production that were chosen for inclusion in this analysis. It provides a brief technical description of each type of plant, the approximate year the technology is expected to become commercially available, optimal nameplate (or rated) capacity, expected capacity factor for operation, resource requirements, and co-products. Those resource requirements that could meet with significant constraints are shown in bold.

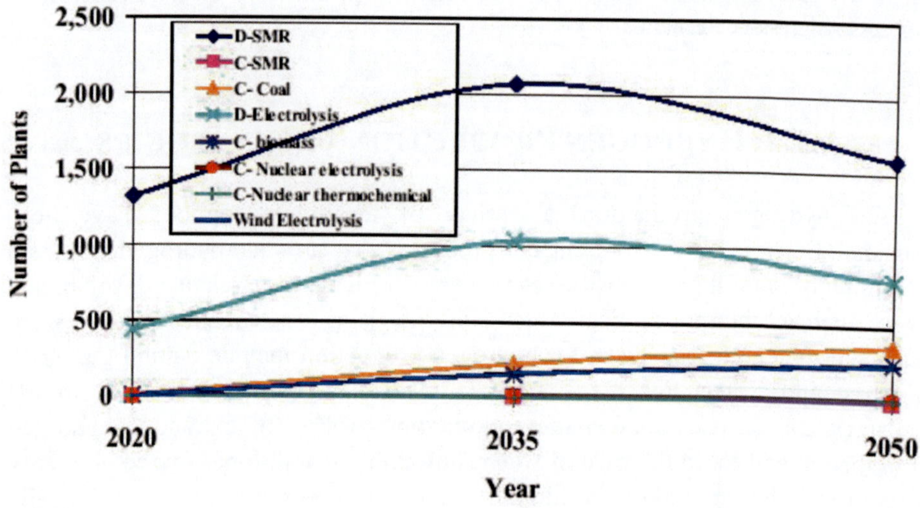

Estimated

Figure A.1.2. Number of Hydrogen Production Facilities by Technology, HFI Scenario

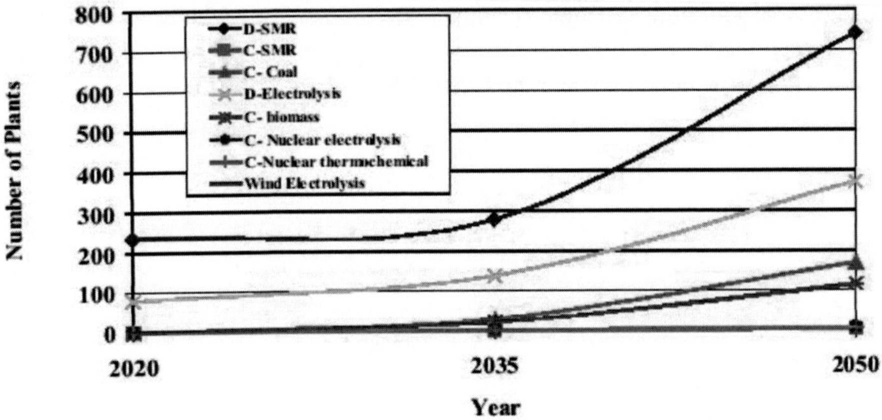

Estimated

Figure A.1.3. Number of Hydrogen Production Facilities by Technology, Less Aggressive Scenario

Table A.1.2. Hydrogen Production Facilities in 2020, 2035 and 2050, by Scenario

	Minimum Number of Units to Satisfy Demand					
	HFI Scenario			Less Aggressive Scenario		
Year	2020	2035	2050	2020	2035	2050
D-SMR, 1,500 kg/d	1319	2073	1586	235	277	741
C-SMR	3	4	4	1	1	2
C-Coal Gasification+CCS	5	224	363	1	30	170
D-Electrolysis, 1,500 kg/d	440	1036	794	79	139	370
C-Biomass	4	156	251	1	21	118
C-Nuclear electrolysis	0	14	13	0	2	7
C-Nuclear thermochemical	0	9	14	0	2	7
C-Wind electrolysis	3	159	243	1	22	114
Estimated						

Due to their different market penetration assumptions, each scenario requires a different quantity of hydrogen production for transportation and stationary uses. Supplying these quantities requires a mix of production facilities which must be built in an orderly fashion over time. Thus, it was necessary to develop plant construction profiles that would result in sufficient hydrogen production capacity being available in the target years of 2020, 2035, and 2050. The resulting technology mixes are shown in Table A.1.2. Figures A.1.2 and A.1.3 show the number of production facilities on-line over time. Although these figures show many distributed production facilities, particularly in the early years of the scenarios, their relatively small capacity limits their contribution to the supply mix.

A.1.3. Hydrogen Delivery Technologies

For hydrogen produced in centralized facilities, delivery to retail stations was assumed to occur in one of the following forms:

Cold compressed gas via truck with tube trailer
Cryogenic liquid via tanker truck
Compressed gas via pipeline

The delivery systems associated with each of these forms are likely to be quite different from one another. High-pressure gases, whether delivered via truck or pipeline, require pressure vessels and compressors; cryogenic liquids require insulated tanks and pumps. Table A.1.4 summarizes the key features of the three hydrogen delivery technologies examined in this study. For this analysis, the H2A delivery model provided detailed cost estimates for each of the components in the three delivery technologies. Component breakdowns were then weighted to reflect the mix of delivery technologies in the scenario and extrapolated to derive national totals.

A.1.4. Feedstock and Resources

Although hydrogen is the tenth most common element on Earth, it is the lightest element, and accounts for less than 1 percent of Earth's total mass. Because it is usually found in compounds (pure hydrogen rarely occurs in nature, although volcanoes and some oil wells release small amounts), energy must be expended to release the hydrogen and, therefore, a feedstock is required for its production. Today, most hydrogen is produced from fossil fuels, primarily natural gas. The natural gas acts as both the feedstock and the energy source. However, for the reasons set forth in sections A.1.4.1 through A.1.4.3, this study assumes that most future hydrogen production will be from coal and biomass.

Table A.1.4. Hydrogen Distribution Technologies Included in this Analysis

Characteristics	Cold Compressed Truck	Liquid Truck	Pipeline
Operating pressure	7,000 psi	100 psi	300-1,000 psi
Unit capacity	1,100 kg/tube trailer	4,000 kg/truck	Diameter and pressure dependent
Components in delivery infrastructure system	Truck cab, tube trailer, terminal, compressors, storage, loading equipment, cooling	Tanker truck, liquefier, terminal, storage, pumps, loading equipment	Compressors, geologic storage, transmission, main and service pipe-line

Source: Technology characterizations from H2A Hydrogen Delivery Scenario Analysis Model (HDSAM) long-term cases.

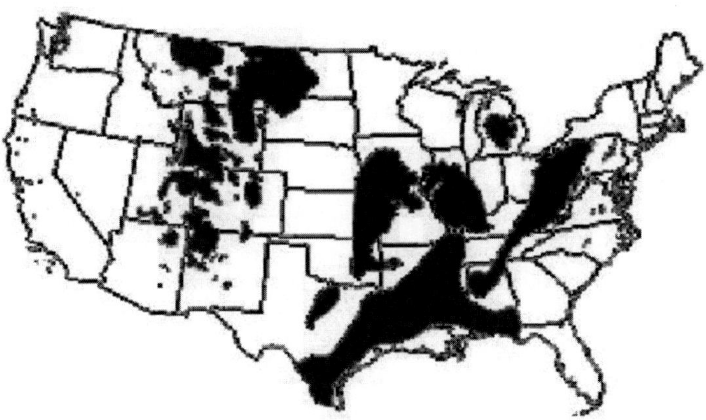

Source: U.S. Geological Survey, accessed September 2006 at http://energy/er.usgs.gov/products/databases/CoalQual/index.htm.

Figure A.1.4. Coal Deposits in the Continental United States

When hydrogen is produced from a hydrocarbon source (such as these), carbon dioxide is also produced. The CO_2 can be captured and used for enhanced oil recovery, sequestered underground, or released. In the case of biomass feedstock, the carbon released does not generally represent a net increase to the global system because the study assumes that future biomass can recapture it.

Alternatively, water can be used as the feedstock for production of hydrogen. The energy required to split water into its component hydrogen and oxygen can be supplied via electricity or high-temperature heat (available from thermonuclear reactions). The electricity can be supplied from the grid (generated by the mix of sources currently employed by the U.S. power industry), or via dedicated power plants built specifically for the purpose of producing hydrogen. Wind and nuclear power are both considered for this purpose, since they do not burn hydrocarbons and therefore do not emit any CO_2.

A.1.4.1. Coal

Feedstock availability and reliance on imports are important issues that must be addressed when planning for the hydrogen economy. Coal is an important domestic feedstock. The U.S. has large coal reserves, widely distributed in two large bands in the Rocky Mountain states and extending northeast from Texas through the Midwest and Appalachian regions. These regions are shown in Figure A.1.4 [7]

According to the Energy Information Administration,

> In the United States, there are vast deposits of coal--more extensive than those of natural gas and petroleum, the other major fossil fuels. Worldwide, coal is the most abundant of the fossil fuels, and its reserves are also the most widely distributed. Estimates of the world's total recoverable reserves of coal in 2002 were about 1,081 billion short tons. The resulting ratio of coal reserves to production exceeds 200 years, meaning that at current rates of production (and no change in reserves), coal reserves could in theory last for another two centuries. The distribution of coal reserves around the world varies notably from that of oil and gas, in that significant reserves are found in the United States and the Former Soviet Union (FSU) but not in the Middle East. The United States, with 26 percent, and the FSU, with 23 percent, account for nearly half of global coal reserves.[8]

Under the HFI Scenario, about 330 million metric tones of coal could be used to produce hydrogen in 2050. This is an increment of one-third over to current U.S. coal consumption, which was just over one billion metric tons in 2005.[9] Using the nation's **our vast domestic coal reserves** to produce hydrogen, is thus a very attractive option. If all the carbon dioxide produced from coal gasification plants operating in 2050 under the HFI Scenario were sequestered over a 30-year lifetime, the total mass to be sequestered would be approximately 34 billion metric tons. Compared to a recent estimate that 98 billion metric tons[10] of storage capacity is available in depleted oil and gas wells, this suggests that sequestration capacity is not likely to be a constraint. However, carbon sequestration technology has not been deployed on a broad scale to date; the stability of the sequestered material, once injected, is a potential challenge; and the availability of sequestration sites of sufficient size to capture needed economies of scale may put a cap on the total usable sequestration capacity. This study is put forward on the express assumption that these issues can and will be resolved in a timely manner to support the transformation to a hydrogen economy.

A.1.4.2. Natural Gas

Natural gas is also under consideration as a feedstock for hydrogen production. **Indeed, it is the primary source for today's production.** However, production of the large quantities of hydrogen required by a hydrogen economy would require significant increases in the supply of natural gas. This could only be met by greatly expanded imports of liquefied natural gas, putting the U.S. energy sector in much the same position of dependence on imports as it is in today, by merely shifting from one form of imports to another. This is not an acceptable option. Thus, the scenarios assume that, although some initial hydrogen is produced from natural gas, use of that feedstock is reduced over time to avoid reliance on imported liquefied natural gas.

A.1.4.3. Biomass

The availability of biomass for production of hydrogen is somewhat uncertain, since estimates depend heavily on assumptions about what land can be used and how much productivity can be improved. For example, Figure A.1.5 shows that 400 million dry tons/year could be produced with moderate yield improvements.[11] About 70% of this material could be from crop residues (cellulosic ethanol, for example). This total could be increased to 600 million t/yr with the use of perennial crops and either further yield improvements or more land dedicated, and to a billion tons with both yield improvements and more land. Under the HFI Scenario, annual demand for biomass feedstock would be about 150 million metric tons, a quantity that could be supplied without significant shifts in land usage.

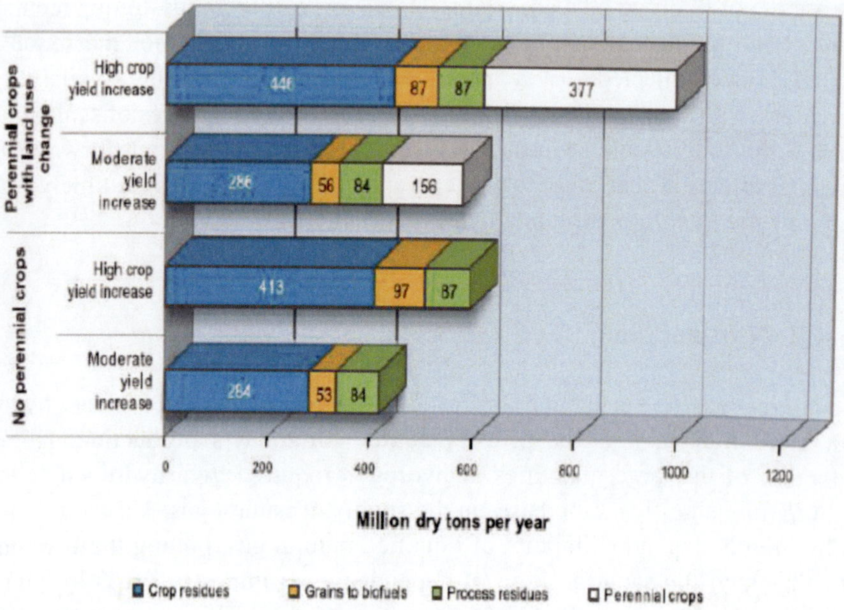

Source: Perlack, R.D., Wright, L., Turhollow, A., Graham, R., Stokes, B., Erbach, D., *Biomass as Feedstock for a Bio-Energy and Bio-Products Industry: The Technical Feasibility of a Billion-Ton Annual Supply*, Oak Ridge National Laboratory for U.S. Department of Energy, April 2005.

Figure A.1.5. Potentially Available Agricultural Resources

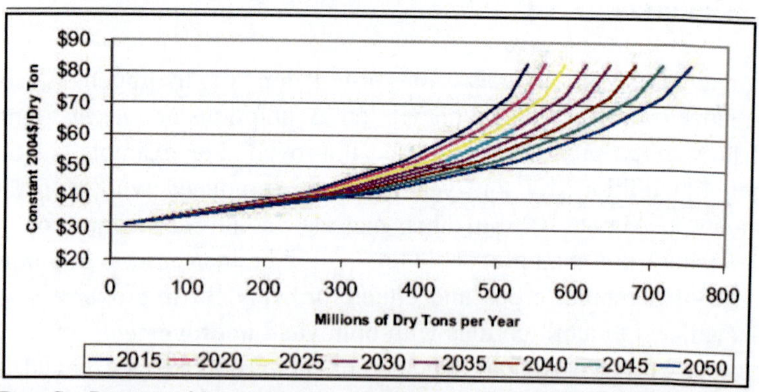

Source: Das, S., *Review of Biomass Supply Curves in EIA and EERE Models*, Oak Ridge National Laboratory, presented to PAE Thursday Analysis Meeting, U.S. Department of Energy, Washington, D.C., May 25, 2006.

Figure A.1.6. Supply of Cellulosic Feedstock as a Function of Cost and Year

Figure A.1.6 shows the estimate for how much cellulosic biomass (the typical crop is hybrid poplar or switchgrass) is likely to be available to produce hydrogen as a function of cost and time.[12] Production of the biomass required for gasification could be accomplished at the assumed price of about $40/dry ton. Since most production is likely to be in the eastern part of the U.S., there will also be regional differences in feedstock transportation costs.[13]

A.1.4.4. Uranium and Water

Nuclear-generated hydrogen has important potential advantages. Nuclear hydrogen requires no imported fossil fuels, results in lower emissions of greenhouse gases and other pollutants, lends itself to large-scale production, and is sustainable. For hydrogen to be produced using nuclear energy, there must be sufficient supplies of uranium (or other fissionable materials) available. It is assumed that, by the time such production processes are commercially available, they will utilize fissionable materials efficiently and produce minimal quantities of low-level waste. Thus, supply will not be a major issue.

Water availability is an important site-specific issue that must be addressed during the siting of hydrogen production plants. Water is required in all hydrogen production processes either as feedstock and/or process water. For example, producing one kilogram of hydrogen by electrolysis requires at least 9 kilograms (about 2.5 gallons) of feedstock water. Similarly, steam reforming of methane requires at least 4.5 kilograms (1.25 gallons) of water, both as feedstock and process water, to produce 1 kilogram of hydrogen gas. The percent of input water returned to the environment depends on the efficiency of the reforming process.

A.1.4.5. Wind

The availability of appropriate sites for wind power generation is constrained by several factors, most notably prevailing wind speed and duration. Figure A.1.7 displays recent estimates of wind resource availability in the contiguous United States.[14] It shows that the most significant resources (those in the highest power-availability classes) are in the Rocky Mountains and Great Plains. Since these areas tend to have limited water availability, siting of some units for the purpose of hydrogen production may be problematic. However, since it has been reported that U.S. wind resources have the potential to produce between 45 and 120 quads of electricity annually, an amount equivalent to between 1,500 and 4,000 GW of electricity generation capacity and well above total current installed U.S. capacity of about 1,000 GW, this may not be a significant constraint. The total wind generating capacity assumed for hydrogen production in 2050 under the HFI scenario is roughly 140 GW, well within the range of available capacity.[15]

A.1.5. STATIONARY FUEL CELLS

In the HFI Scenario, it is assumed that stationary fuel cells will provide 1% of incremental electricity generation over 2015 levels in 2020, 5% of incremental generation over 2015 in 2035, and 10% of the incremental generation in 2050. Assuming 70% capacity utilization of the fuel cells, these investment costs are estimated at $110 million in 2020, $670 million in 2035, and $3.3 billion in 2050. In the Less Aggressive Scenario, although stationary fuel cells do not come into widespread use until after 2020, they provide 2% of incremental electricity generation in 2035 and 5% in 2050, with corresponding investment costs of $270 million in 2035 and $1.67 billion in 2050.

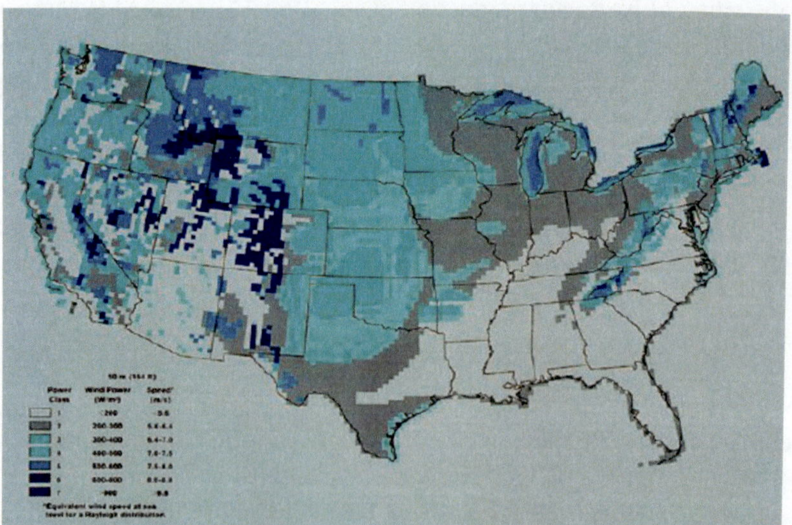

Source: National Renewable Energy Laboratory, *Wind Energy Resource Atlas of the United States*, accessed September 2006 at http://rredc.nrel.gov/wind/pubs/atlas/maps/chap2/2-06m.html.

Figure A.1.7. U.S. Wind Resources by Wind-Speed Class

Figure A.1.8 shows how the construction costs for stationary fuel cells depend on the capacity factor at which they are utilized.[16] This is particularly important, because many of the early installations of stationary fuel cells will be as backup units where a power outage is particularly critical, such as hospitals or computer centers. These fuel cells will only be used when the primary source of power (the grid) goes down, which is expected to occur less than 1% of the time (3-4 days per year). The capacity factor for these cells may be very low. By contrast, cells in remote installations would be utilized more, as would residential and commercial units. For this study, we have assumed an average capacity factor of 75%.

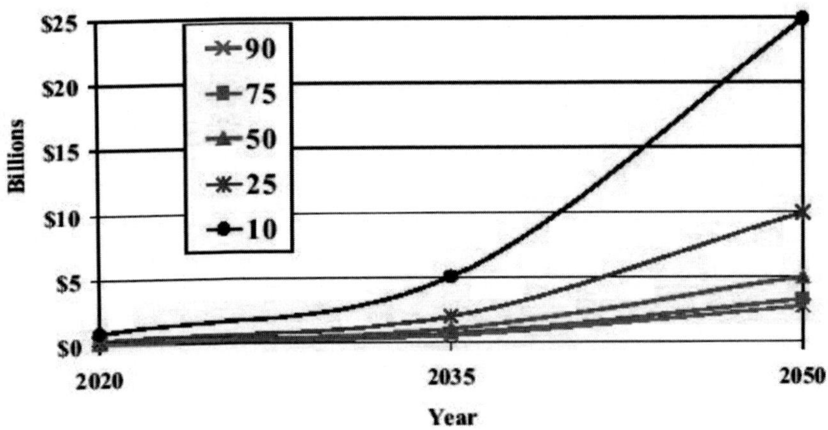

Estimated

Figure A.1.8. Annual Cost of Stationary Fuel Cell Construction as a Function of Capacity Factor (%) of Existing Units

End Notes

[1] Singh, M., Vyas A., Steiner E., *VISION Model: Description of Model Used to Estimate the Impact of Highway Vehicles, Technologies and Fuels on Energy Use and Carbon Emissions to 2050*, Argonne National Laboratory Report ANL/ESD/04-1, Dec. 2003 and subsequent updates, accessed August 2006 at http://www.transportation.anl.gov/pdfs/TA/299.pdf.

[2] *Hydrogen Posture Plan*, U.S. Department of Energy report, February 2004, accessed August-September 2006 at http://www1.eere.energy.gov/hydrogenandfuelcells/posture_plan04.html.

[3] *30x30: A Scenario for Supplying 30% of 2004 Motor Gasoline with Ethanol by 2030*, Appendix H – Final Lifecycle Environmental Analysis of Biofuels: GREET and VISION Simulation Results, U.S. Department of Energy Office of Biomass Programs draft report, July 2006.

[4] *Hydrogen Posture Plan*, U.S. Department of Energy report, February 2004, accessed August-September 2006 at http://www1.eere.energy.gov/hydrogenandfuelcells/posture_plan04.html.

[5] *FY 2007 GPRA Benefits Estimates*, U.S. Department of Energy report in response to requirements of the Government Performance and Results Act, accessed August-September 2006 at http://www1.eere.energy.gov/ba/pba/2007_benefits.html.

[6] Forecourt design generally refers to fueling station facility and equipment design.

[7] U.S. Geological Survey, accessed September 2006 http://energy/er.usgs.gov/products/databases/CoalQual/index.htm.

[8] U.S. Department of Energy, Energy Information Administration, accessed September 2006 at http://www.eia.doe.gov/neic/infosheets/coal.

[9] U.S. Department of Energy, Energy Information Administration U.S. Coal Consumption by End-Use Sector, accessed September 2006 at http://www.eia.doe.gov/cneaf/coal.

[10] Stevens, S., Kuuskraa, V., Taber, J., *CO_2 Sequestration in Depleted Oil and Natural Gas Fields*, IEA Greenhouse Gas R&D Program, 1999.

[11] Perlack, R.D., Wright, L., Turhollow, A., Graham, R., Stokes, B., Erbach, D., *Biomass as Feedstock for a Bio-Energy and Bio-Products Industry: The Technical Feasibility of a Billion-Ton Annual Supply*, Oak Ridge National Laboratory for U.S. Department of Energy, April 2005.

[12] Das, S. *Review of Biomass Supply Curves in EIA and EERE Models*, Oak Ridge National Laboratory, presented to PAE Thursday Analysis Meeting, U.S. Department of Energy, Washington, D.C., May 25, 2006.

[13] The eastern part of the United States is defined as the western boundary of the first line of states just west of the Mississippi River. The source of this definition was from an email note from Robert D. Perlack to Donald Jones of RCF, Inc.

[14] National Renewable Energy Laboratory, *Wind Energy Resource Atlas of the United States*, accessed September 2006 at http://rredc.nrel.gov/wind/pubs/atlas/maps/chap2/2-06m.html.

[15] Elliott, D.L., Schwartz, M.N., *Wind Energy Potential in the United States*, Pacific Northwest National Laboratory report PNL-SA-23109, NTIS no. DE94001667, September 1993.

[16] Capacity factor is the ratio of annual operating hours to total hours (8,760 per year), generally expressed as a percentage.

APPENDIX 2. INFRASTRUCTURE COST ESTIMATION

Appendix 1 provides a description of VISION and sets forth the technology assumptions used to estimate hydrogen production and delivery infrastructure requirments. This appendix (Appendix 2) provides an overview of the H2A models (which provided key inputs to infrastructure cost estimation) and the cost estimation process itself, as well as detailed tables of infrastructure costs that were used to construct the hydrogen fuel industry vector for IMPLAN.

In order to model employment impacts, hydrogen production and delivery infrastructure were characterized in sufficient detail to permit development of a hydrogen fuel industry vector for the IMPLAN input-output model. That vector requires inputs into (purchases from) each of 509 economic sectors and outputs from (sales to) each of those sectors. Two separate types of purchases are modeled – one corresponding to capital costs of infrastructure (which are heavily oriented toward equipment and construction sectors) and another corresponding to operating costs (which include feedstock, energy, labor, and other production inputs).

Several models were used to develop estimates of capital and operating costs of infrastructure. DOE's series of H2A models provided detailed cost breakdowns for individual production and delivery technologies. As needed, these were supplemented by off-line analyses, as well as with the VISION model. The VISION forecasts of vehicle sales, stocks, and fuel use were used to extrapolate H2A results for individual units to national aggregates.

A.2.1. H2A MODELS

Research and development of hydrogen production and delivery technologies and their associated infrastructure require a robust, comprehensive set of analytic tools. Not only must such tools yield clear results to guide R&D decisions, but their methodologies must be objective, rigorous and open to oversight and review. Over the past several years. DOE's Office of Hydrogen, Fuel Cells and Infrastructure Technologies (OHFCIT) has supported the development of a suite of tools collectively known as H2A (or Hydrogen Analysis) Models. These tools supplied many of the inputs to this study.

The H2A project was initiated in response to a plethora of hydrogen production and delivery infrastructure analyses, many of which contained findings that appeared inconsistent or were otherwise, questionable. Discrepancies resulted from differences in analytical bases (e.g., whether the analyses were based on current or advanced technologies, on targets or empirical results, on "real world' or simulated duty cycles, etc.), as well as in the many economic, financial, and technological assumptions used in the analyses. Beginning in 2002, H2A sought to establish a consistent set of financial parameters and methodologies for cost analyses. The goal was to improve the transparency of hydrogen analysis so researchers could better understand similarities and differences among efforts and so the industry could better validate results. To that end, OHFCIT leveraged the combined talents and capabilities of analysts from several national laboratories, universities, and the private sector to develop a single work plan and uniform a set of analytic tools.

Initially, the HFCIT analysts identified the following objectives for H2A:

1. Establish a standard format for modeling and reporting results of production, delivery, and forecourt (refueling station) cost analyses
2. Identify specific parameters for modeling production, delivery, and forecourt
3. Seek better validation of methodologies and results through continued dialogue with industry
4. Improve how demand is factored into supply/infrastructure analysis
5. Establish a mechanism for rapid dissemination of methodologies and results
6. Identify gaps in hydrogen production and delivery analyses

In order to better focus the required effort, the group of analysts was divided into three teams – centralized production, distributed production, and delivery. Within their respective subject areas, each of these teams was directed to focus their activities on items 1-6 above.

The centralized production team was charged with developing a spreadsheet tool and analysis of hydrogen production from fossil fuels (gasification and reforming of coal and natural gas, with or without carbon sequestration), nuclear power (thermo-chemical water-splitting in a high temperature, gas-cooled reactor with a Sulfur Iodine cycle), and renewable fuels (wind and biomass gasification). The assumptions discussed in Appendix 1 regarding hydrogen production are incorporated in the centralized production team's analysis. The resulting tool is now available at http://www.hydrogen.energy.gov/h2a_production.html.

The distributed, or "forecourt" production team was charged with developing a similar spreadsheet tool for two sizes of distributed production facilities – 100 kg/d and 1,500 kg/d – using steam methane reforming and electrolysis technologies. That tool has been combined with the first of several tools developed by the hydrogen delivery team and is also available at the above address.

The hydrogen delivery team, was charged with developing tools to model the cost contribution of all activities between the centralized production of hydrogen and its use on-board a vehicle. Two tools have been developed – the Delivery Components Model (which incorporates the distributed production tools mentioned above) and the Delivery Scenarios Model. Version 1.0 of both of these models and their documentation are also available at the above web address.

A.2.1.1. H2A Production and Forecourt Models

The H2A production and forecourt models provide a standardized, transparent technique for comparing alternative financial and technological assumptions and for generating detailed breakdowns of capital and operating costs for a production facility of given process design, output (nameplate capacity and capacity factor), and level of technology development. Table A.2.1 lists the hydrogen production technologies for which H2A models have been developed, their key parameters, and potential co-products. These latter factors are discussed further in Appendix 1.

Table A.2.1. Description and Key Features of Hydrogen Production Technologies Included in this Study

Technology	Description	Capacity T/d (CF%)	Co-products
(C-SMR) Centralized steam methane reforming, with sequestration	Natural gas feedstock (450,000 lb//hr @750° F and 450 psia) is de-sulfurized, mixed with process stea-m and undergoes strongly endothermic reac-tions over Ni-based catalyst at 1400-1700°F. Pro-cess gas (CO & H_2) passes through a heat recov-ery step and a water gas shift reactor to produce additional H_2.	380 (90)	Steam
(D-SMR) Distributed steam meth-ane reforming	Hydrodesulfurized natural gas is steam reformed at 10-atm with PSA gas cleanup. Factory built, skid-mounted unit is assumed to achieve 75% H H_2 recovery. H_2 exits PSA at 300 psi and is com-pressed and stored at 6250 psi for cascade filling.	1.5 (70)	None
(D-Electroly-sis) Distribut-ed electrolysis	H2 from high pressure (300 psi) alkaline electro-lysis is compressed to 6250 psi. Process excludes CO_2 capture & co-product sale at this time.	1.5 (70)	None
(C-Coal Gas-if) Central-ized coal gas-ification, with carbon seque-stration	Coal is fed into a commercially-available gasifier (Conoco-Phillips EGas), and the product gas is sent to conventional gas cooling, shift conversion, acid gas cleanup (where byproduct sulfur is produced), and PSA purification. The CO_2 is compressed to 2200 psi for sequestration.	246 (90)	Electricity , elemental sulfur

(C-Biomass) Mid-size bio-mass gasific-ation, no seq-uestration	Woody biomass (represented as hybrid poplar) is fed to a gasifier (indirectly heated with hot sand), conventional catalytic steam reformer, high- and low-temperature water gas shift reactors, and PSA. Steam is the fluidizing gas; no oxygen (or air) is fed to the gasifier.	164 (90)	None
(C-Wind) Mid-size wind electro-lysis	A 276 MW wind farm with 185 x 1.5 MW wind turbines and 119 electrolyzers produces an aver-age of 50,000 kg/d H_2. The installation sells about 324 X 106 kWh of byproduct power back to the grid annually. No O_2 is assumed to be sold.	86 (58)	Electricity
(C- Nuclear TC) Central-ized nuclear thermochem-ical water splitting	Advanced high-temperature gas-cooled reactors (HTGR) (4 x 600 MWt modules) provide direct process heat for the sulfur-iodine thermo-chemical process for splitting of water.	768 (90)	O_2
(C-Nuclear Elec) Centralized nuc-lear electrolysis	Nuclear power from HTGR is used to generate electricity, some of which is used to split water and some of which is sold to the grid. Co-product O_2 can also be sold.	719 (90)	Electr-icity, O_2

Source: H2A models.

A.2.1.2. H2A Delivery Model

Also known as the Hydrogen Delivery Scenario Analysis Model (HDSAM), the H2A delivery model estimates the cost of delivering hydrogen from a centralized production facility to a hydrogen-fueled vehicle. Like other H2A-developed tools, HDSAM is Excel-based and uses an engineering economics approach to cost estimation. For a given scenario, a set of components (e.g., compressors, tanks, tube-trailers, etc.) is specified, sized and linked into a simulated delivery system or pathway. Financial, economic and technological assumptions are then used to compute the cost of those components and their overall contribution to the delivered cost of hydrogen. HDSAM Version 1.0 contains many default values that represent currently available (2005) technologies and costs. For this report, a modified version of the model incorporating the DOE program's 2017 cost targets was used.

Hydrogen delivery is defined as the entire process of moving hydrogen from the gate of a centralized production plant to its point of consumption in a light-duty vehicle. Thus, delivery includes all transport, storage, and conditioning activities from the outlet of a centralized hydrogen production facility to a refueling station that stores, dispenses, and, in some cases, further conditions the hydrogen fuel. In HDSAM, user selection of a delivery mode invokes an associated chain of delivery components or processes. For example, if the user selects liquid hydrogen truck delivery for a given market and selects a specific assumed hydrogen vehicle penetration rate,[1] the model calculates not only the number and cost of the trucks required to deliver the fuel to refueling stations, but also the cost of appropriately-sized liquefiers, terminal and depot storage, truck loading facilities, and refueling stations.

Figure A.2.1 illustrates the three delivery options modeled. For this study, the cost of all loading, conditioning, and storage activities normally associated with a terminal or depot, as well as costs for storage at refueling stations, are combined with production costs in the hydrogen fuel industry vector developed for the IMPLAN model. Section A.2.2.2 provides further discussion of delivery.

A.2.2. Cost Estimation

As stated above, two sets of costs were estimated for the input-output analysis – annual costs of producing hydrogen fuel (which are based on hydrogen demand by light-duty vehicles and discussed in Appendix 1) and infrastructure construction costs which are incurred over several years in response to actual and anticipated growth in the demand for hydrogen fuel. Cost estimation required the following steps:

Developing unit cost data
Regrouping production and delivery components
Aggregating unit costs into appropriate categories
Estimating total demand for hydrogen production and delivery

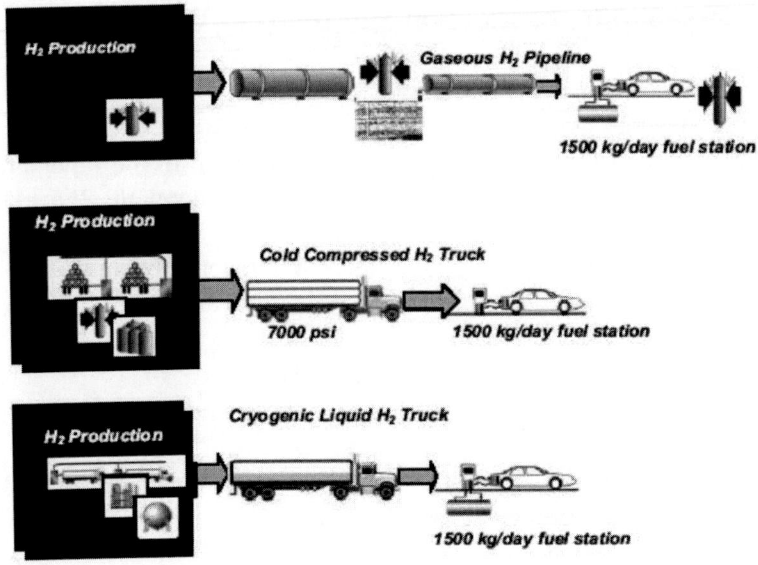

Figure A.2.1. Delivery Pathways Included in Cost Characterization

A.2.2.1. Development of Unit Cost Data

Since outputs from the H2A models are not in a form that can be directly used for the input-output analysis, several steps were required to develop the necessary information.

A separate H2A run was analyzed for each technology. Each run includes a detailed process schematic, complete with stream flows, for the optimum scale based on best engineering judgment. In the case of delivery options, the scale is a user input. Based on the mix of centralized production facilities and their sizes (based on the assumptions set forth in Appendix 1), a weighted-average hydrogen-production plant size of roughly 250,000 kg/d was assumed for the delivery runs. For the production runs, capital equipment and construction costs, annual fixed operating and maintenance costs, variable operating costs (e.g., for feedstock and energy) that depend on actual facility utilization, and capital replacement costs with a known schedule were estimated from the default process design. These costs were summarized to create unit costs for each component. Miscellaneous costs (e.g., marketing costs, debt service, road taxes, sales taxes, and value-added profit) were then added to estimate unit sales prices of the final product.

A.2.2.2. Regrouping Production and Delivery Components

The production and delivery components contained in the H2A models were regrouped to separate hydrogen refueling (or retailing), pipeline distribution, and trucking (both liquid and cold compressed hydrogen) from the other production and delivery components. The remaining delivery components (i.e., liquefiers, terminals, and all forms of storage throughout the production and delivery chain) were proportionally combined (based upon hydrogen demand) with central hydrogen production costs. The resulting combinations and their components are:

Distributed H2 Production – SMR (without Forecourt)
Distributed H2 Production – Electrolysis (Grid Electricity) (without Forecourt)
Central H2 Production – SMR plus Terminal and Storage
Central H2 Production – Coal Gasification plus Terminal and Storage
Central H2 Production – Biomass Gasification plus Terminal and Storage
Central H2 Production – Nuclear Electrolysis plus Terminal and Storage
Central H2 Production – Nuclear Thermo-chemical plus Terminal and Storage
Central H2 Production – Wind Electrolysis plus Terminal and Storage
Delivery – H2 Pipeline System
Delivery – Trucking (a combination of liquid and cold gas truck)
Retail Forecourt

Costs estimated for each of these combinations included production capital expenditures, replacement capital, fixed operating and maintenance costs, feedstock costs, energy costs, other raw material costs, water costs, and other variable operating costs.

Table A.2.2. U.S. Centralized Hydrogen Production: Demand and Capital Costs (2050 HFI Scenario)

FEEDSTOCK	Natural Gas	Coal	Biomass	Uranium	Uranium	Wind Power
TYPE	N. America		Energy Crop	Nuclear	Nuclear	Class 6
TECHNOLOGY	NG SR	Gasification +CCS	Gasification[a]	Electrolysis	Thermo-chemical	Electrolysis
CATEGORY	Central Plant plus Terminal	Central Plant plus Terminal	Central Plant plus Terminal	Central Plant plus Terminal	Central Plant plus Terminal	Central Plant plus Terminal
H2 Capacity & Demand						
Rated capacity (tonne H2/day/unit)	379.4	246.5	163.6	719.4	768	101.5
Capacity Factor	90%	90%	90%	90%	90%	50%
H2 Throughput (tonne H2/day/unit)	341.5	221.9	147.2	647.5	691.2	50.2
Total Units in U.S.	2	400	228	12	13	501
Year	2050	2050	2050	2050	2050	2050
U.S. Demand (% shares)	0.50%	53.00%	20.00%	4.50%	5.50%	15.00%
U.S. Demand (k tonne H2/day)	0.8	88.8	33.5	7.5	9.2	25.1
U.S. H2 Production Capital (10^6\$)						
Equipment, Turbine/Generator	\$0	\$28,307	\$0	\$0	\$0	\$76,995
Equipment, Electrical	\$15	\$11,716	\$594	\$4,920	\$865	\$445
Equipment, Compressor	\$171	\$56,133	\$7,502	\$1,311	\$1,602	\$4,369
Equipment, Pump	\$14	\$1,480	\$559	\$126	\$154	\$419
Equipment, Dispensing/Metering	\$0	\$8	\$3	\$1	\$1	\$2

Table A2.2. (Continued)

FEEDSTOCK	Natural Gas	Coal	Biomass	Uranium	Uranium	Wind Power
TYPE	N. America		Energy Crop	Nuclear	Nuclear	Class 6
TECHNOLOGY	NG SR	Gasification +CCS	Gasification[a]	Electrolysis	Thermo-chemical	Electrolysis
CATEGORY	Central Plant plus Terminal	Central Plant plus Terminal	Central Plant plus Terminal	Central Plant plus Terminal	Central Plant plus Terminal	Central Plant plus Terminal
Equipment, Reactor	$189	$20,237	$5,147	$19,370	$5,255	$20,273
Equipment, Pressure Vessels	$108	$11,457	$4,323	$973	$6,018	$3,243
Equipment, Other Machinery	$51	$15,027	$2,362	$861	$2,787	$20,503
Equipment, Trucks	$0	$0	$0	$0	$0	$0
Equipment, Pipeline	$0	$7	$3	$1	$540	$2
Process Chemicals	$0	$0	$0	$0	$1,378	$0
Installation, Construction	$151	$38,662	$16,069	$678	$6,291	$10,371
Installation, Steel Structure	$0	$2,538	$1,368	$0	$1	$1
Installation, Land Costs	$21	$2,274	$915	$216	$263	$785
Installation, Site Work	$27	$4,286	$1,296	$239	$292	$7,792
Installation, Other	$47	$34,370	$7,016	$425	$520	$2,530
Management, Contingency, Fees	$44	$41,333	$7,573	$393	$4,378	$11,566

[a] An indirect heat gasification process, chosen for the H2A advanced biomass gasification case, is used in this analysis.

Table A.2.3. U.S. Distributed Hydrogen Production & Delivery: Demand and Capital Costs (2050 HFI Scenario)

FEEDSTOCK	Natural Gas	Electricity	H2	H2	H2
TYPE	N. America	Grid Mix			
TECHNOLOGY	NG SR	Electrolysis	Pipeline System	Truck	
CATEGORY	Forecourt Production w/o Retail	Forecourt Production w/o Retail	Delivery	Delivery	Retail Forecourt
H2 Capacity & Demand					
Rated capacity (tonne H2/day/unit)	1.5	1.5	277.8	277.78	1.500
Capacity Factor	70%	70%	90%	90%	70%
H2 Throughput (tonne H2/day/unit)	1.1	1.1	250.0	250.00	1.050
Total Units in U.S.	1,596	798	469	191	157,227
Year	2050	2050	2050	2050	2050
U.S. Demand (% shares)	1.0%	0.5%	70.0%	28.5%	98.5%
U.S. Demand (k tonne H2/day)	1.7	0.8	117.3	47.8	165.1
Total U.S. H2 Production Capital (10^6\$)					
Equipment, Turbine/Generator	\$0	\$0	\$0	\$0	\$0
Equipment, Electrical	\$30	\$15	\$0	\$0	\$0
Equipment, Compressor	\$537	\$228	\$0	\$0	\$0
Equipment, Pump	\$16	\$0	\$0	\$0	\$0
Equipment, Dispensing/Metering	\$0	\$0	\$0	\$0	\$10,727
Equipment, Reactor	\$717	\$478	\$0	\$0	\$0

FEEDSTOCK	Natural Gas	Electricity	H2	H2	H2
TYPE	N. America	Grid Mix			
TECHNOLOGY	NG SR	Electrolysis	Pipeline System	Truck	
CATEGORY	Forecourt Production w/o Retail	Forecourt Production w/o Retail	Delivery	Delivery	Retail Forecourt
Equipment, Pressure Vessels	$615	$308	$0	$0	$0
Equipment, Other Machinery	$621	$478	$7,868	$0	$0
Equipment, Trucks	$0	$0	$0	$17,809	$0
Equipment, Pipeline	$0	$0	$8,216	$0	$0
Process Chemicals	$0	$0	$0	$0	$0
Installation, Construction	$260	$154	$38,160	$0	$6,385
Installation, Steel Structure	$0	$0	$0	$0	$0
Installation, Land Costs	$0	$0	$6,434	$0	$0
Installation, Site Work	$103	$51	$0	$0	$1,596
Installation, Other	$80	$40	$0	$0	$1,596
Management, Contingency, Fees	$261	$157	$0	$0	$3,512

Source: Based on H2A analysis.

Table A.2.4. U.S. Centralized Hydrogen Production Costs (2050 HFI Scenario)

FEEDSTOCK	Natural Gas	Coal	Biomass	Uranium	Uranium	Wind Power
TYPE	N. America		Energy Crop	Nuclear	Nuclear	Class 6
TECHNOLOGY	NG SR	Gasification +CCS	Gasification[a]	Electrolysis	Thermo-chemical	Electrolysis
CATEGORY	Central Plant plus Terminal	Central Plant plus Terminal	Central Plant plus Terminal	Central Plant plus Terminal	Central Plant plus Terminal	Central Plant plus Terminal
U.S. H2 Production (10^6\$/year)						
Replacement Capital						
Equipment, Turbine/Generator	$0	$0	$0	$0	$0	$705
Equipment, Electrical	$0	$0	$0	$144	$79	$0
Equipment, Compressor	$15	$1,544	$582	$131	$160	$437
Equipment, Pump	$1	$74	$28	$6	$8	$21
Equipment, Dispensing/Metering	$0	$0	$0	$0	$0	$0
Equipment, Reactor	$0	$0	$104	$0	$0	$661
Equipment, Pressure Vessels	$5	$573	$216	$49	$59	$162
Equipment, Other Machinery	$0	$17	$111	$162	$81	$666
Equipment, Trucks	$0	$0	$0	$0	$0	$0
Fixed O&M Costs						
Labor cost	$6	$4,708	$1,216	$366	$426	$260
Fringe on all labor, excluding VHSL	$3	$2,744	$709	$213	$248	$152
Overhead labor excluding Fringe, VHSL	$3	$2,679	$687	$208	$242	$139
Soft Costs (w/o rent, insurance, taxes)	$8	$7,003	$1,941	$757	$743	$1,779

FEEDSTOCK	Natural Gas	Coal	Biomass	Uranium	Uranium	Wind Power
TYPE	N. America		Energy Crop	Nuclear	Nuclear	Class 6
TECHNOLOGY	NG SR	Gasification +CCS	Gasification[a]	Electrolysis	Thermo-chemical	Electrolysis
CATEGORY	Central Plant plus Terminal	Central Plant plus Terminal	Central Plant plus Terminal	Central Plant plus Terminal	Central Plant plus Terminal	Central Plant plus Terminal
Debt Service	$99	$32,015	$6,520	$2,283	$2,389	$11,872
Business Tax	$0	$1,947	$501	$0	$0	$34
Value Added, Profit	$0	$4,544	$1,170	$1	$1	$80
Property Taxes and Insurance	$19	$5,402	$1,111	$594	$611	$3,198
Rent	$5	$569	$215	$48	$59	$161
Material Costs for Maintenance & Repairs	$9	$1,613	$394	$383	$488	$1,667
Other Fixed O&M	$5	$411	$155	$147	$173	$1,137
Sales and Road Tax	$176	$18,677	$7,048	$1,586	$1,938	$5,286
Feedstock and Energy						
Feedstock and Equipment Transport						
Coal	$0	$10,000	$0	$0	$0	$0
Natural Gas	$414	$0	$734	$0	$0	$0
Diesel (Petroleum)	$0	$0	$0	$0	$3	$0
Uranium 3% U232	$0	$0	$0	$979	$1,121	$0
Biomass Energy Crop	$0	$0	$6,124	$0	$0	$0
Grid Power	$89	$6,531	$3,116	$554	$1,003	$1,848
Water	$5	$161	$27	$41	$40	$45

FEEDSTOCK	Natural Gas	Coal	Biomass	Uranium	Uranium	Wind Power
TYPE	**N. America**		**Energy Crop**	**Nuclear**	**Nuclear**	**Class 6**
TECHNOLOGY	**NG SR**	**Gasification +CCS**	**Gasification[a]**	**Electrolysis**	**Thermo-chemical**	**Electrolysis**
CATEGORY	**Central Plant plus Terminal**	**Central Plant plus Terminal**	**Central Plant plus Terminal**	**Central Plant plus Terminal**	**Central Plant plus Terminal**	**Central Plant plus Terminal**
Other Raw Materials						
Electrolyte	$0	$0	$0	$0	$0	$401
Inert Gas	$0	$0	$0	$0	$0	$5
Process chemicals	$11	$0	$1,623	$0	$0	$0
Water (non-feedstock)	$0	$0	$77	$0	$0	$214
Other Variable Operating Costs	$0	$771	$439	$0	$0	$0

[a] An indirect heat gasification process, chosen for the H2A advanced biomass gasification case, is used in this analysis.

Table A.2.5. U.S. Hydrogen Distributed Production, Delivery, and Retail Costs (2050 HFI Scenario)

FEEDSTOCK	Natural Gas	Electricity	H2	H2	H2
TYPE	N. America	Grid Mix			
TECHNOLOGY	NG SR	Electrolysis	Pipeline System	Truck	
CATEGORY	Forecourt Production w/o Retail	Forecourt Production w/o Retail	Delivery	Delivery	Retail Forecourt
U.S. H2 Production (10^6\$/year)					
Replacement Capital					
Equipment, Turbine/Generator	$0	$0	$0	$0	$0
Equipment, Electrical	$0	$0	$0	$0	$0
Equipment, Compressor	$0	$0	$0	$0	$0
Equipment, Pump	$0	$0	$0	$0	$0
Equipment, Dispensing/Metering	$2	$1	$0	$0	$536
Equipment, Reactor	$32	$17	$0	$0	$0
Equipment, Pressure Vessels	$0	$0	$0	$0	$0
Equipment, Other Machinery	$32	$17	$197	$0	$0
Equipment, Trucks	$0	$0	$0	$890	$0
Fixed O&M Costs					
Labor cost	$0	$0	$49	$1,549	$2,690
Fringe on all labor, excluding VHSL	$0	$0	$29	$903	$1,567
Overhead labor excluding Fringe, VHSL	$11	$6	$8	$261	$452
Soft Costs (w/o rent, insurance, taxes)	$83	$43	$15	$466	$810
Debt Service	$215	$128	$6,068	$2,137	$2,858
Business Tax	$6	$1	$0	$0	$0
Value Added, Profit	$14	$3	$1	$0	$1

FEEDSTOCK	Natural Gas	Electricity	H2	H2	H2
TYPE	N. America	Grid Mix			
TECHNOLOGY	NG SR	Electrolysis	Pipeline System	Truck	
CATEGORY	Forecourt Production w/o Retail	Forecourt Production w/o Retail	Delivery	Delivery	Retail Forecourt
Rent	$40	$20	$0	$0	$2,873
Material Costs for Maintenance & Repairs	$101	$62	$303	$161	$479
Other Fixed O&M	$0	$0	$29	$930	$160
Sales and Road Tax	$352	$176	$0	$0	$0
Feedstock and Energy					
Feedstock and Equipment Transport					
Coal	$0	$0	$0	$0	$0
Natural Gas	$884	$0	$0	$0	$0
Diesel (Petroleum)	$0	$0	$0	$3,681	$0
Uranium 3% U232	$0	$0	$0	$0	$0
Biomass Energy Crop	$0	$0	$0	$0	$0
Grid Power	$181	$861	$0	$0	$0
Water	$0	$4	$0	$0	$0
Other Raw Materials					
Electrolyte	$0	$0	$0	$0	$0
Inert Gas	$0	$0	$0	$0	$0
Process chemicals	$0	$0	$0	$0	$0
Water (non-feedstock)	$6	$7	$0	$0	$0
Other Variable Operating Costs	$3	$1	$0	$0	$0

Source: Based on H2A analysis.

Table A.2.6. Per Unit Capital Costs for Hydrogen Distributed and Centralized Plant Production

FEEDSTOCK	Biomass	Electricity	Uranium	Wind Power
TYPE	Energy Crop	Nuclear	Nuclear	Class 6
TECHNOLOGY	Gasification[a]	Electrolysis	Thermo-chemical	Electrolysis
CATEGORY	Central Plant	Central Plant	Central Plant	Central Plant
H2 Production Total Capital (10^6\$)				
Equipment, Turbine/Generator				\$153.82
Equipment, Electrical		\$410.91	\$52.64	
Equipment, Compressor	\$7.37			
Equipment, Pump				
Equipment, Dispensing/Metering				
Equipment, Reactor	\$22.61	\$1,662.84	\$394.01	\$40.50
Equipment, Pressure Vessels			\$362.08	
Equipment, Other Machinery	\$9.03	\$68.02	\$202.64	\$40.50
Equipment, Pipeline			\$40.41	
Process Chemicals			\$103.33	
Installation, Construction	\$57.34		\$409.56	\$16.20
Installation, Steel Structure	\$6.00			
Installation, Land Costs	\$0.25	\$2.00	\$2.00	\$0.28
Installation, Site Work	\$1.02			\$13.97
Installation, Other	\$22.52			\$2.22
Management, Contingency, Fees	\$25.59		\$292.27	\$20.49
Replacement Capital (10^6\$/y)				
Equipment, Turbine/Generator				\$1.41
Equipment, Electrical		\$12.40	\$5.92	
Equipment, Dispensing/Metering				
Equipment, Reactor	\$0.46			\$1.32
Equipment, Other Machinery	\$0.46	\$13.80	\$5.92	\$1.32

[a] An indirect heat gasification process, chosen for the H2A advanced biomass gasification case, is used in this analysis.

Source: Based on H2A analysis

A.2.2.3. Aggregation of Unit Costs

In order to create input vectors for use in the IMPLAN runs, it was necessary to take costs from the H2A model runs for each production and delivery combination (see above) and allocate them to the appropriate sectors in the 509-sector IMPLAN model. This process required considerable judgment, as there is not a one-to-one correspondence between cost elements and economic sectors. Once the allocation was made, the cost elements for each technology included in the scenario being analyzed were multiplied by the assumed participation of that technology in the scenario's mix, to get a total cost vector that was appropriately weighted.

A.2.2.4. Estimating Total Demand from Hydrogen Production and Delivery

Table A.2.2 shows total capital costs for centralized hydrogen production plants (plus terminal) based upon the hydrogen throughput per plant and number of plants (which are also shown in the table) for the various feedstock combinations. In Table A.2.2, as well as in all tables and figures presented below, values shown are for the HFI Scenario for the year 2050. While similar values were calculated for the years 2020 and 2035, as well as for the Less Aggressive Scenario, they are not included here. Table A.2.3 shows comparable detail for distributed hydrogen production (forecourt reforming and electrolysis, but excluding hydrogen retailing), for pipeline and truck delivery, and for hydrogen retailing. Production costs include replacement capital, fixed operating and maintenance (O&M) costs, taxes, feedstock costs, energy costs, raw materials costs, and other variable operating costs.

A.2.3. Total U.S. Infrastructure Cost Details

Tables A.2.4 through A.2.7 contain total costs (in millions of dollars) for each of the production-and-delivery combinations for the HFI Scenario (aggressive hydrogen market expansion). Figure A.2.2 and A.2.3 show unit costs ($/kg of hydrogen) broken down for each individual component. Figure A.2.2 displays capital costs while Figure A.2.3 displays fixed O&M and variable costs. As a point of comparison, the figures also show similar cost elements for gasoline.

Figure A.2.2 shows estimated hydrogen unit cost by cost category for each production and delivery technology. Delivered cost is the sum of centralized production, delivery, and retail forecourt costs. For distributed production, delivered cost is the sum of distributed reforming of natural gas or distributed electrolysis and retail forecourt cost.

Figure A.2.3 shows the distribution of the unit cost of hydrogen (including delivery and retail forecourt) by production technology. These values were derived from total U.S. costs divided by hydrogen demand (by production technology) in 2050 under the HFI Scenario. The Less Aggressive Scenario yields similar results because it is based on the same production and infrastructure unit costs.

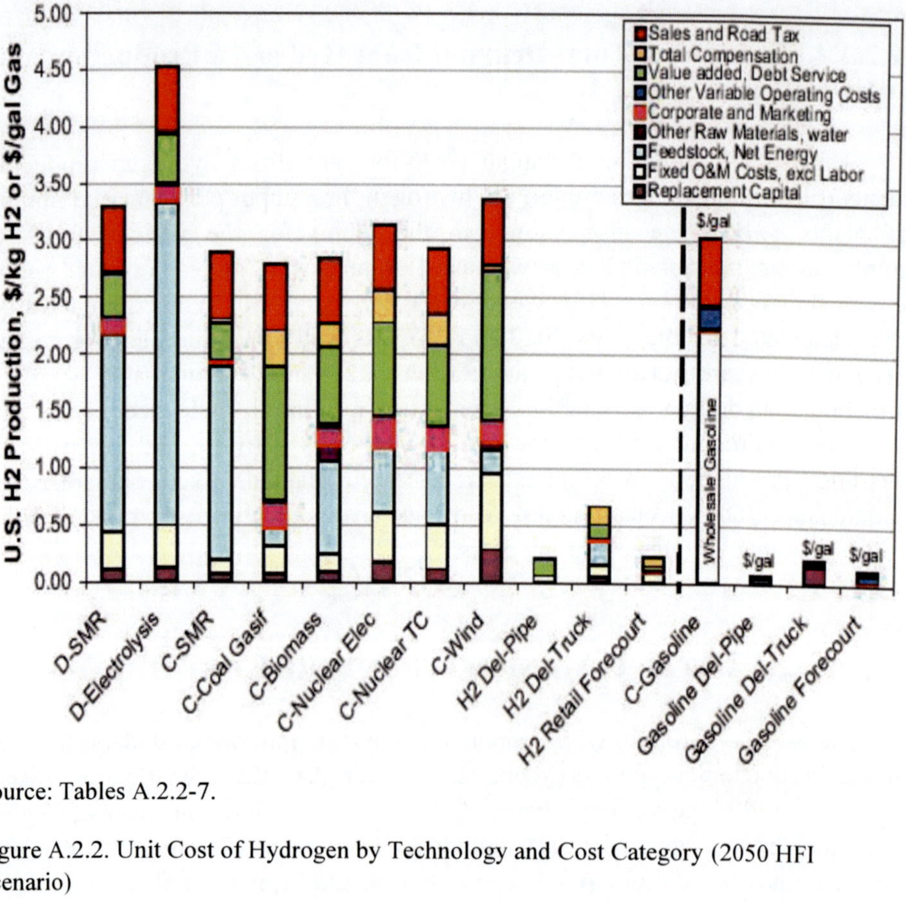

Source: Tables A.2.2-7.

Figure A.2.2. Unit Cost of Hydrogen by Technology and Cost Category (2050 HFI Scenario)

Table A.2.7. Per Unit Fixed O&M and Variable Costs for Hydrogen Distributed and Centralized Plant Production

FEEDSTOCK	Natural Gas	Electricity	Natural Gas	Coal	Biomass	Uranium	Uranium	Wind Power
TYPE	N. America	Grid Mix	N. America		Energy Crop	Nuclear	Nuclear	Class 6
TECHNOLOGY	NG SR	Electrolysis	NG SR	Gasification +CCS	Gasification[a]	Electrolysis	Thermo-chemical	Electrolysis
CATEGORY	Forecourt Production	Forecourt Production	Central Plant	Central Plant	Central Plant	Central Plant	Central Plant	Central Plant
Fixed O&M Costs ($/year)								
Labor cost	17,057	17,057	2,000,000	11,520,000	5,184,000	30,720,000	31,200,000	465,600
Fringe on all labor, excluding VHSL	9,940	9,940	1,165,468	6,713,094	3,020,892	17,901,583	18,181,295	271,321
Overhead labor excluding Fringe, VHSL	9,846	9,846	1,154,532	6,650,106	2,992,548	17,733,617	18,010,705	268,775
Soft Costs (w/o rent, insurance, taxes)	57,012	59,552	3,024,000	17,418,240	8,478,165	64,789,228	55,461,996	3,537,917
Debt Service	152,467	177,853	16,248,120	64,235,314	18,207,005	150,063,900	130,124,960	20,158,703
Business Tax	3,670	1,835	2,294	4,863,392	2,202,291	20,646	25,235	68,822
Value Added, Profit	8,564	4,282	5,353	11,347,914	5,138,678	48,175	58,881	160,584
Property Taxes and Insurance	43,562	50,815	3,623,818	10,710,000	3,034,501	42,875,348	37,178,595	5,759,629
Rent	43,194	43,194						
Material Costs: Maintenance & Repairs	66,322	80,704	811,810	2,195,000	511,740	27,557,300	30,891,925	2,916,018
Other Fixed O&M	1,000	1,000	520,000			9,637,200	9,762,000	2,040,000
Sales and Road Tax	220,774	220,774	71,795,725	46,646,405	30,959,967	136,134,648	145,332,410	10,560,084

FEEDSTOCK	Natural Gas	Electricity	Natural Gas	Coal	Biomass	Uranium	Uranium	Wind Power
TYPE	**N. America**	**Grid Mix**	**N. America**		**Energy Crop**	**Nuclear**	**Nuclear**	**Class 6**
TECHNOLOGY	**NG SR**	**Electrolysis**	**NG SR**	**Gasification +CCS**	**Gasification[a]**	**Electrolysis**	**Thermo-chemical**	**Electrolysis**
CATEGORY	**Forecourt Production**	**Forecourt Production**	**Central Plant**	**Central Plant**	**Central Plant**	**Central Plant**	**Central Plant**	**Central Plant**
Variable Costs ($/year)								
Feedstock and Energy								
Coal				24,976,156				
Natural Gas	553,963		168,521,240		3,225,096			
Diesel (Petroleum)							230,347	
Uranium 3% U232							84,076,553	84,076,553
Biomass Energy Crop						26,902,315		
Grid Power	113,273	1,078,752	11,285,198		2,860,588		24,380,897	
Water		5,625	2,089,274	401,860	118,238	3,484,415	2,975,196	89,728
Other Raw Materials								
Electrolyte								800,848
Inert Gas								9,381
Process chemicals			4,629,000		7,130,000			
Water (non-feedstock)	3,825	8,946	14,813		338,591			428,093
Other Variable Operating Costs	1,800	1,800		1,925,000	1,930,000			

[a] An indirect heat gasification process, chosen for the H2A advanced biomass gasification case, is used in this analysis.

Source: Based on H2A analysis.

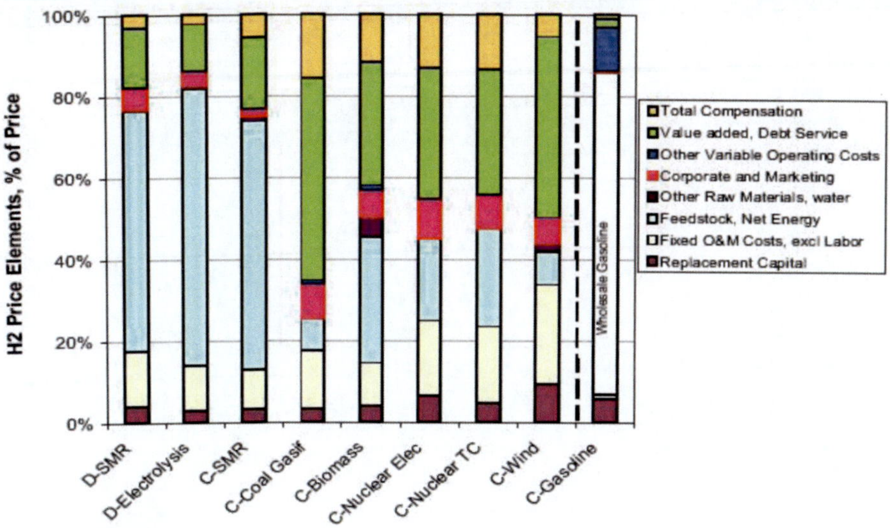

Source: Tables A.2.2-7.

Figure A.2.3. Distribution of Unit Fuel Cost (2050 HFI Scenario)

End Notes

[1] The scale and, therefore, the cost of facilities needed to effect delivery turns on a number of factors, including the delivery method and the assumed vehicle penetration rate.

APPENDIX 3. VEHICLE COST ESTIMATES

This Appendix develops the cost structure of hydrogen vehicles by component, then assigns those component costs to the industries that supply them in order to develop a new inter-industry purchase vector for hydrogen vehicle production. This purchase vector, which is added to the IMPLAN inter-industry matrix, will be used to estimate the direct and indirect labor requirements for supplying the economy with this new technology. The structure of this Appendix is therefore parallel to that of Appendix 2, which modified the IMPLAN purchase vectors to account for the infrastructure investments required to implement the hydrogen economy.

The confluence of issues surrounding energy security and cost, air quality, and global warming have led to major government initiatives in fuel cells and alternative energy sources. Evolving government regulations, with the states such as California often taking the lead, are placing increasingly stringent environmental (i.e., emissions), efficiency (i.e., fleet mileage or CAFÉ), and Greenhouse gas (i.e., CO_2) requirements original equipment manufacturers (car companies or OEMs). As these regulations evolve, they will continue to stimulate innovation in powertrain technologies, thereby challenging engineers to meet the above requirements while maintaining or, preferably, improving performance and meeting market price expectations.

Recent tensions in the Middle East combined with the energy demands of China primarily and India have led to increases in oil prices and concerns about United States energy security. Due to the large contribution of the transportation sector to oil demands, the government has directed the Department of Energy (DOE) to look for long-term solutions to the nation's growing dependence on imported energy sources, particularly oil. Since the early 1970s, the government has initiated a series of programs, generally in concert with the OEMs, to develop alternatives to internal combustion engines. For example, the U.S. Advanced Battery Consortium (USABC) was established to advance the introduction of electric vehicles by reducing battery prices through development of low-cost manufacturing technologies. The Partnership for a New Generation of Vehicles (PNGV) was organized to develop the technologies needed to introduce fuel cells, light weight materials, and advanced power electronics into more fuel efficient low emissions vehicles. With the announcement of the FreedomCAR program, the PNGV's focus shifted to developing the technologies needed to implement a hydrogen based transportation sector. While the PNGV focused on a reformate fuel cell system to utilize the existing gasoline distribution network, the FreedomCAR program redirected efforts to a hydrogen-based fuel infrastructure and long-term utilization of renewable and domestic energy sources. Major efforts have been initiated to develop production, delivery, and storage (on-board) technologies for hydrogen.

The automotive industry has continued to advance internal combustion engine technologies (both gasoline and diesel) to reduce emissions and to, improve efficiency and performance. Advances in diesel technology for light-duty vehicles, i.e., passenger cars and light duty trucks, has led to the dominance of diesel powertrains in Europe, while Toyota's development and highly successful introduction of the hybrid Prius coincided with the recent rise in oil prices. Both hybrid and diesel technologies improve fuel economy. However, hybridization of diesel powertrains will depend on the overall economies (i.e., life cycle cost). As fuel costs increase and emission standards become more stringent, internal combustion engine developers still have many options to further improve that technology along with advances in hybrid technology.[1,2]

Overall vehicle design (glider[3] and powertrain) will also play an important role in fuel economy and vehicle cost. The advances in powertrain technology over the past twenty years have gone into performance (acceleration and ride) rather than improving vehicle fuel economy.

Increasing electrification of powertrains, auxiliary loads, and cabin amenities will continue. Fuel cells provide a technology path to satisfy the power and range requirements of electric vehicles. The oil shocks of the 70's led to electric vehicle programs, but the available battery technologies could not meet the vehicle range, life, and cost competition set by internal combustion engine vehicles. The investments in electric vehicle powertrains in these programs did advance electric component technologies that have benefited current hybrid and fuel cell vehicles, such as motors and power electronics. Other related activities have developed technologies to reduce parasitic power losses from belt driven auxiliary loads, such as drive-by-wire, brake-by-wire, and motor driven air conditioning systems.

Radiator fans have been electric for many years. Additionally, consumer demands for digital systems in the cabin, i.e. navigation and entertainment, will continue to increase cabin power demands. Advances in battery technology, i.e. nickel metal hydride (NiMH) and the shallow discharge demands of hybrid electric vehicles have led to acceptable battery life (e.g., >8 years) in vehicles such as the Prius. Developers and OEMs are pursuing Lithium-ion technology in order to improve energy and power density for both Prius-like hybrids and more energy-intensive plug-in hybrids.

The PNGV and FreedomCar programs have led to significant advances in fuel cell vehicles, as evidenced by the vehicles in DOE's ongoing demonstration and validation project. This project is collecting data on fuel cell life in a variety of ambient conditions. Honda has leased a fuel cell vehicle to a family in California. Life and cost of fuel cell and hydrogen storage systems continue to be additional factors of importance for long-term commercial viability. To address hydrogen infrastructure issues, the FreedomCAR program has increased R&D efforts in hydrogen production, delivery, and storage.

For this report, the development of manufacturing and cost projections for fuel cell and conventional vehicle powertrains has relied heavily on the experience of the team in particular projects conducted for EPRI[4,5] and the DOE.[6,7,8] The EPRI project looked at the economics of various hybrid scenarios relative to conventional vehicles, while projects for the DOE have assessed the cost of fuel cell systems, the impact of various fuel chains on cost, pollution, and CO_2 emissions,[7,8] and other work involving hybrid technologies and drive cycle analysis. The newer EPRI study extended the hybrid analysis to hybrid SUVs that fall under the category of light-duty trucks.

A.3.1. Powertrain Technology Options

Table A.3.1 illustrates how powertrains and fuel type may evolve over time leading to a transformation from conventional internal combustion engine powered vehicles to fuel cell vehicles. With time, hybridization increases electrification of the powertrain, while fuel flexible internal combustion engine vehicles will operate in a marketplace with a mix of fuels (petroleum, biofuel, and hydrogen) as a hydrogen infrastructure is developed. Examples include E85 vehicles which operate on ethanol and gasoline and BMW's hydrogen or gasoline fueled internal combustion engine vehicle. It is assumed that the fuel cell vehicle will be hybridized; because a hybrid engine will operate on a mix of fuels, hybridization will permit a reduction in the size of the fuel cell and provide transient (ability to switch from one power source to another as needed) performance. Additionally, increasing the battery size for operation as a plug-in hybrid further improves mileage and allows reduction of the hydrogen storage tank. Currently, hybrids are being introduced into the market place and the batteries needed for plug-in hybrids are currently under development.

A.3.2. FUEL CELL SYSTEM TECHNOLOGY DESCRIPTION

Since William Grove's discovery of the fuel cell principle in 1839, many variations of the technology have been considered. However, the first practical application of fuel cells was the Apollo missions of the 1960's when alkaline fuel cells provided electrical power for NASA space missions. Subsequently, developers have researched phosphoric acid (PAFC), molten carbonate (MCFC), and solid oxide (SOFC) fuel cells for a number of stationary applications, including but not limited to centralized and distributed electrical power generation. For example, United Technologies has installed 270 PureCell 200 (200 kW) units in 19 countries and 85 cities, in a variety of applications (e.g., reliable power for banking operations, distributed generation, and on military bases).[9] More recently, Fuel Cell Energy has been installing units in applications involving distributed generation and installations using landfill gas. Purchase and operation of all of these applications have been subsidized by government funds. In the mid-90's, stimulated by advances in polymer membrane fuel cells (PEMFCs) and increasing reliance on uncertain and more expensive oil imports DOE initiated a major effort in concert with the car companies to develop automotive fuel cell powertrains by 2015.

Fuel cells are electrochemical systems with the characteristics of both batteries and chemical reactors. In a primary battery, the amount of reactants contained within the can determines the available energy or run-time. In rechargeable batteries, these reactants can be restored by plugging into the wall, allowing multiple uses. In a fuel cell, the reactants, e.g., hydrogen and oxygen (air), are supplied from external sources, and the run-time is determined by the size of the reservoirs for these chemicals. Figure A.3.1 illustrates the conceptual configuration of a fuel cell. The critical components are the electrodes where the fuel (hydrogen) is oxidized at the anode and the oxidant (oxygen in air) is reduced at the cathode.

The electrolyte, in contact with both electrodes, provides ionic continuity between the anode and cathode. The fuel types mentioned in the previous paragraph are differentiated by several characteristics, but the names are associated with the type of electrolyte (phosphoric acid, molten carbonate, alkaline aqueous solution, and proton conducting polymer membranes). The use of these electrolytes leads to different temperatures of operation (from ambient to 1000 °C) and electrode materials. As shown in Figure A.3.1, the net reaction is the same as chemically burning hydrogen and oxygen, but with the benefit of producing electricity directly from the reaction. Benefits of fuel cells include:

Higher theoretical efficiency than heat engines, leading to more efficient use of fuels

Zero emissions of pollutants and CO_2 at the tailpipe when fueled with hydrogen

Table A.3.1. Powertrain Configuration Options

Powertrain Type	**Fuel Storage**	**Powertrain**				
		Primary Power		**Electric**		
Conventional Internal Combustion Engine (ICE)	Molded Tank for gasoline	ICE	Transmission			PbAc SLI Battery
Hybrid Electric Vehicle	Molded Tank for gasoline	Smaller ICE	Smaller Transmission	Electric Motor/ Generator	Power Electronics	NiMH or Li-ion Battery
Plug-in Hybrid	Molded Tank for gasoline	Smaller ICE	Smaller Transmission	Electric Motor/ Generator	Power Electronics	Li-ion Battery
Hydrogen ICE Hybrid Electric Vehicle	Compressed H2 Storage	Smaller ICE	Smaller Transmission	Electric Motor/ Generator	Power Electronics	NiMH or Li-ion Battery
Fuel Cell Hybrid Electric Vehicle	Compressed H2 Storage	Fuel Cell Stack	Transmission	Electric Motor/ Generator	Power Electronics	Li-ion Battery
Fuel Cell Plug-in Hybrid Electric Vehicle	Compressed H2 Storage	Fuel Cell Stack	Transmission	Electric Motor/ Generator	Power Electronics	Li-ion Battery

Source: determined in this study.

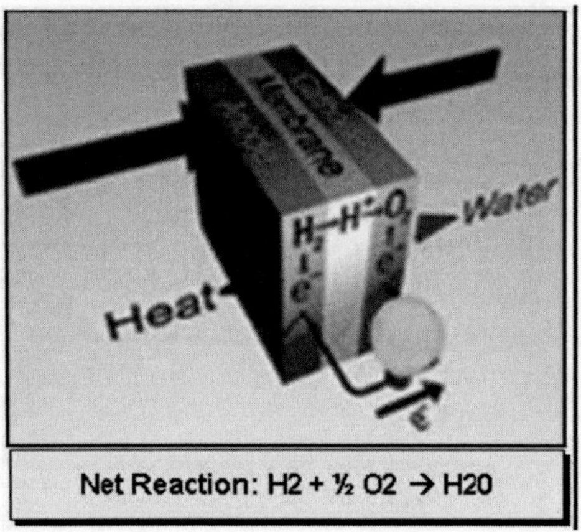

Source: Fuel Cells Today.

Figure A.3.1. Reaction Occurring in a Fuel Cell

In practice, fuel efficiency, pollutant emissions, and CO_2 emissions of a fuel cell vehicle must be evaluated on a well-to-wheels[10] basis to obtain a true assessment of these metrics relative to existing technologies.[7, 8,11]

An automotive PEMFC is shown in Figure A.3.2 to illustrate how the individual fuel cell elements are electrically connected in series to form a high voltage (e.g. 200 – 300 volts). The stack also contains passages for distributing fuel, air, and coolant to the appropriate layers. An individual (unit) fuel cell layer (membrane electrode assembly) consists of several layers, including:

Electrolyte in the form of proton exchange membrane

Anode and cathode electrodes with platinum catalysts to promote the fuel and air reactions

Gas diffusion layers for each electrode to uniformly distribute the gaseous reactants to each electrode

Bipolar plates electrically connect each unit cell in series to form the stack. Each bipolar plate provides several additional functions, including flow channels for fuel and air, cooling for each cell, and isolation of adjacent fuel and air streams.

The stack must then be assembled into a system that provides air, fuel, thermal and water management, and overall controls of all of the subsystems. An example of a prototype fuel cell subsystem without hydrogen storage is shown in Figure A.3.3.

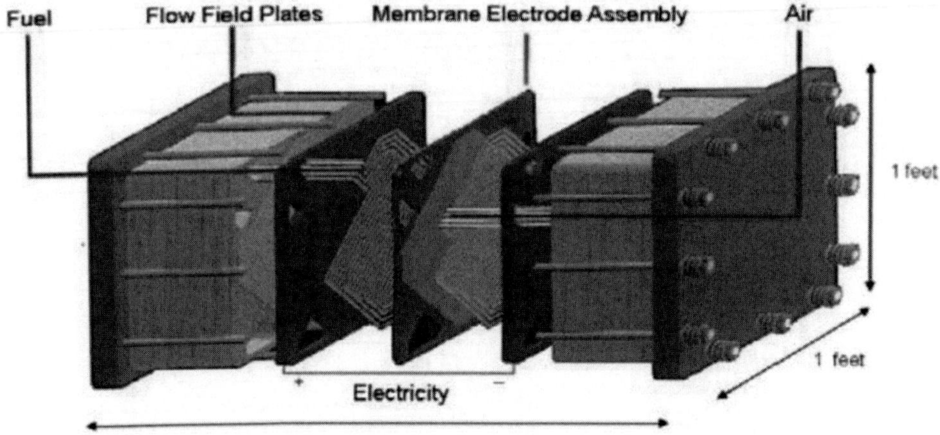

Source: Ballard.

Figure A.3.2. Automotive PEMFC Stack, Showing Series Connection of Cells to Form a Stack

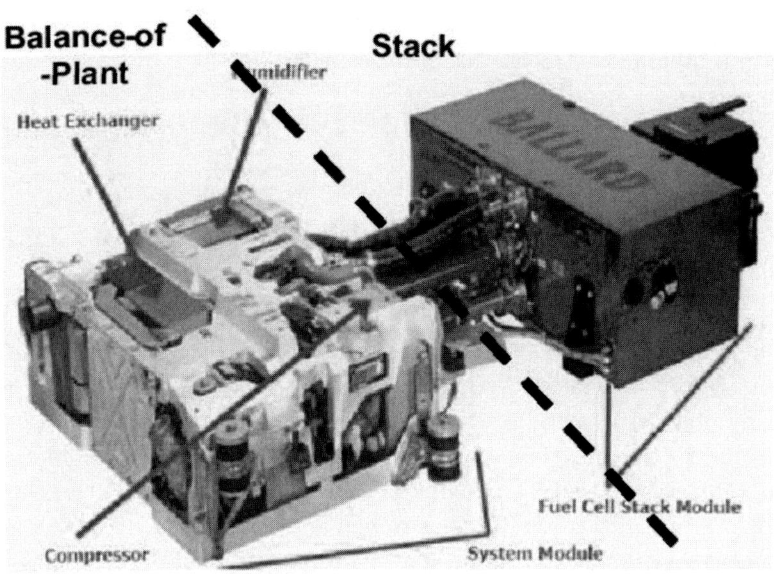

Source: Ballard.

Figure A.3.3. Prototype Fuel Cell Subsystem, Showing Packaging of an Automotive Stack and Balance-of-Plant Components

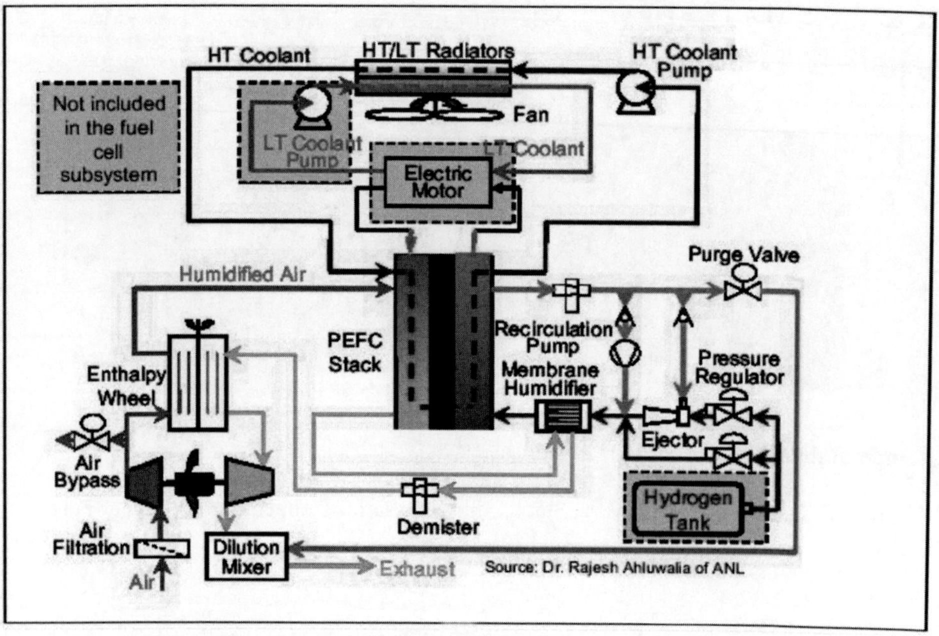

Source: Dr. R. Ahluwalia, Argonne National Laboratory

Figure A.3.4. Overall System Configuration with Major Components

A.3.3. 2005 Baseline Fuel Cell System Cost Analysis

Year 2005 technology and cost projections were used as a baseline for the fuel cell system assessment.[6] The costs in this study were based on 2005 technology, but estimated as if the technology was made at high volumes (500,000 units per year). Figure A.3.4 shows the fuel cell system configuration used in this assessment, including: the stack, hydrogen storage, water management (humidification of the hydrogen and air and water recovery from the exhaust air), thermal management, hydrogen recirculation, and air management (pressurization and filtration of the inlet air and recovery of energy from the exhaust air). A cost of $108/kW was projected for the 80-kW system with the contributions from the stack, balance-of-plant, and assembly shown in Figure A.3.5.[12] Figure A.3.6 further details the cost of the fuel cell stack.

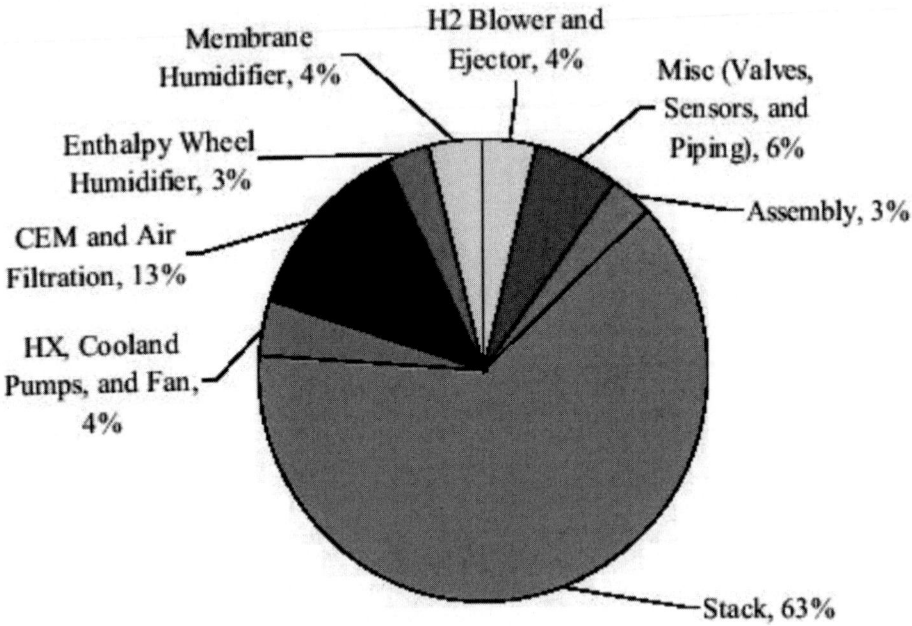

Estimated

Figure A.3.5. Breakdown in Stack and BOP Component Cost Contributions for and 80 kW direct Hydrogen Fuel Cell System ($108/kW, $8,640)

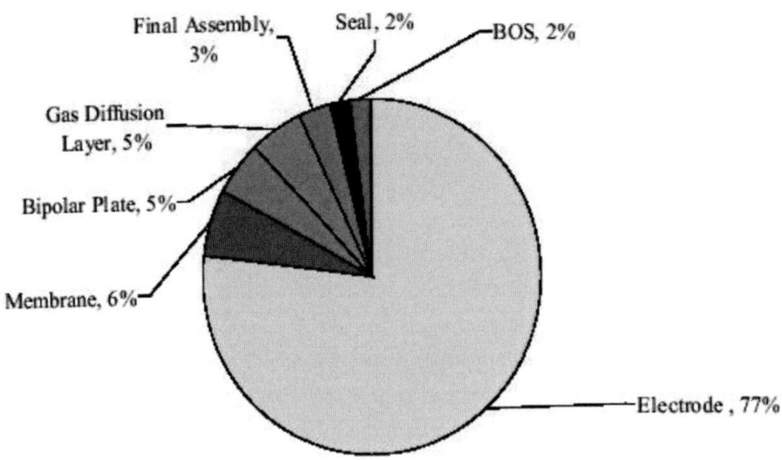

Estimated

Figure A.3.6. Cost Breakdown for a 2005 80kW Direct Hydrogen Stack ($67/kW, $5,360)

Table A.3.2. Assumed Values of Key Cost Drivers

Parameter	2005[a]	Future[b]
Platinum price ($/ troy ounce)	900	900
Total platinum loading (mg/cm^2)	0.75	0.2
Power density (mW/cm^2)	600	1000
Stack Cost ($/kW)	67	19

Sources: [a]TIAX LLC, *Cost Analysis of PEMFC Systems for Transportation*, NREL/SR-560-39104, December 2005; [b]assumptions reported in Section A.3.5.1.

The major system cost drivers from this assessment are:

Power density
Cost of platinum
Platinum loading

In general, fuel cells are a materials-intensive technology. Consequently the amount of power that can be generated per unit area of the electrode and material costs are dominant cost drivers. Power density (power per unit area of the electrode) determines the total amount of material in the stack. Platinum is the most expensive material in the stack. Consequently its cost on a weight basis ($/g) and loading (mg/cm^2) are critical cost drivers. The first column of Table A.3.2 lists the 2005 values assumed for these key drivers in the 2005 study. The DOE goals require developers to increase power density while reducing platinum loading. Traditionally, the price of platinum has averaged approximately $450/troy ounce with fluctuations up and down depending on world events. In the last 20 years, the introduction of catalytic converters in cars to control tailpipe emissions and the development of a jewelry market for platinum created new markets that required significant increases in production capacity. These new markets led to some spikes in platinum price, but prices still fluctuated around historical values. However, in the last five years, the economic prosperity of China has significantly increased the prices of commodities, including platinum. Instability in the Middle East and high oil prices have put further upward price pressure on precious metals, including platinum and gold. As a consequence, platinum prices have remained above a $1,000/troy ounce for the past year. The assumed price, $900/tr oz, is weighted toward the more recent high values.

From a system perspective, in 2005, the proportional allocation between stack and balance-of-plant components was approximately 65/35. As shown in Figure A.3.6, balance-of-plant components contribute approximately 35 percent of the cost. Humidification, pressurization, and hydrogen circulation components contribute to the cost and complexity of the system.

The manufacture of fuel cell stacks will involve different materials and manufacturing processes than internal combustion engines. Internal combustion engines are mainly metal components made by casting, forging, and machining processes. In contrast, a fuel cell stack is comprised of polymer and carbon components fabricated with film coating, paper making, and molding processes. Balance-of-plant components are generally comprised of traditional internal combustion engine system components (i.e., heat exchangers, compressors or blowers, tubes and valves, sensors, and controllers) even though they will be tailored to the special requirements of fuel cells. Humidification of gas streams may entail new technologies involving polymer-based materials.

Table A.3.3 lists the major materials used in the stack and their assembly processes. The electrodes are the most expensive component, but contribute negligible weight and volume to the stack. The bipolar plates are the largest contributor to the mass and volume of the stack.

Table A.3.3. Summary of Fuel Cell Stack Materials and Manufacturing Processes

Component	Material	Process
Electrodes	Platinum catalyst in a coating ink	Screen printing or continuous film coating process
Gas Diffusion Layer	Carbon fiber in a non-woven paper (or woven cloth) with surfactant and carbon coatings	Wet lay paper making process (or a weaving process). Tradit-ional coating processes
Bipolar Plates	Graphite powder and polymer resin Alternative: Stainless steel or other metal with surface treatment or coating	Graphite and resin mixed and molded in a finished part Metal forming process (stamping) with vacuum coating process
Electrolyte	Proton exchange membrane - (e.g. Nafion®) perfluoropolymer with sulfonic acid groups	Coating lines modified to form membranes

A.3.4. 2006 Employment Study Inputs

Power Train Specifications. The first step in developing vehicle and component pricing was to specify the powertrain components and energy storage capacities. For this study, only two powertrain options were considered, conventional and hybrid fuel cell vehicles. Within these two powertrain options, two vehicle types were characterized—a mid-sized passenger car and a sport utility vehicle (SUV)—to represent the range of autos and light trucks of interest. The series hybrid internal combustion engine configuration is shown as an example of how the battery can be used to down-size the engine. The EPRI vehicle specifications[4,5] for conventional and hybrid internal combustion engine vehicles fueled by gasoline were used for the baseline vehicles and as a starting point for the comparable fuel cell vehicles. Table A.3.4 shows the values selected for the conventional and fuel cell vehicles. For the mid-size passenger (light-duty) fuel cell vehicle the electric traction motor was scaled to the fuel cell power assuming it could also handle short duration power pulses up to 120 kW. For the SUV (or light-duty truck), the electric traction motor was specified to handle the combined power output of the fuel cell and battery for longer duration operation at or near peak power.

In this project, technical specifications for conventional internal combustion engine and fuel cell powertrains were developed to provide a basis for identifying changes in technology that would lead to changes in manufacturing manpower. The following paragraphs discuss the assumptions for specifying the future fuel cell system.

A.3.5. Sub-System Costs

Since the fuel cell, hydrogen storage, and battery represent the largest changes in the powertrain, the focused discussion is on these components. Furthermore, DOE goals have been used to project performance and cost in the future systems.

Table A.3.4. Vehicle Powertrain Specifications

Powertrain and Fuel Configuration	Units	Mid-Size Vehicle (LDV)			SUV (Light-Duty Truck, LDT)		
		ICE[a] (Baseline)	Series HEV[b]	Hydrogen Fuel Cell HEV[b]	ICE[a] (Baseline)	Series HEV[b]	Hydrogen Fuel Cell HEV[b]
Vehicle Designation		CV[c]	HEV[b]	FCHEV[d]	CV[c]	HEV[b]	FCHEV[d]
Fuel		Gasoline	Gasoline	cH_2[e]	Gasoline	Gasoline	cH_2[e]
Engine/Fuel Cell Type		V-6[f] DISI[g]	V-6[f] DISI[g]	PEMFC[h]	V-6[g] DISI[f]	V-6[f] DISI[g]	PEMFC[h]
Hybrid Battery		No	Yes	Yes	No	Yes	Yes
ICE Power	kW	27	67		212	145	
Fuel Cell Power	kW			80			140
Battery Power	kW		49	40		80	65
Battery Energy	kWh		2.91	2.91		4.6	4.6
Electric Traction Motor Power	kW		44.3	80.0		65.3	205
Hydrogen Storage Capacity	kg		4.6				6.4

[a] Internal combustion engine; [b] hybrid electric vehicle; [c] conventional vehicle; [d] fuel cell hybrid electric vehicle; [e] compressed hydrogen storage; [f] 6-cylinder internal combustion engine in V geometry; [g] Direct injection spark ignition (engine); [h] proton exchange membrane fuel cell.

Source: estimated in this study.

A.3.5.1. Fuel Cell Sub-system

The 2005 fuel cell system is used as the starting point for projecting the 2020 stack and balance-of-plant components.

A.3.5.1.1. Stack Cost

The 2005 fuel cell cost projection was based on large production volumes (500,000 units per year) and high volume material price estimates. Consequently, in this project, key performance metrics—power density and platinum loading—were revised, based on DOE 2015 goals, and the 2005 material price estimates were used. The second column of Table A.3.2 compares these parameters with the 2005 values. It was elected to not change the platinum price, given the uncertainty in predicting the future prices. The value assumed, $900/tr.oz, is more than twice the 100-year average price, yet less than the $1,325 peak in 2006. In using the $900/tr.oz. price figure, the analysis adopts a conservative view of the cost of fuel cells; the actual realized price over the time period studied will likely on average be less than the estimate. Utilizing the assumed price for platinum, in conjunction with the DOE targets for platinum loading and power density, results in a stack cost of $19/kW, 27% higher than the DOE 2015 stack cost target of $15/kW.

Along with the stack performance improvements, it is assumed that future stacks will utilize higher temperature membrane technology that can operate without humidification of the fuel and air streams. This has significant implications for simplification of the balance-of-plant components related to water management (humidification). Furthermore, it is assumed that the compressor expander can be manufactured at DOE's projected values. The simplification and cost reduction in balance-of-plant components lead to a decrease in that cost from $41/kW in 2005 to $14/kW.

A.3.5.1.2. Stack Materials and Cost

Table A.3.5 and Table A.3.6 show the amount of major stack materials and their cost contributions to the stack in mid-size vehicles and SUVs (80 and 140 kW stack respectively).

Table A.3.5. Stack Material Amounts in Mid-Size Vehicles and SUVs

Stack Material	Units	Material Amount	
		Light-Duty Vehicle	Light-Duty Truck
Membrane (Nafion)	g	809	1,416
Electrode Catalyst (Pt)	g	16	28
Gas Diffusion Layer (Carbon Cloth)	kg	3	6
Bipolar Plate			
Graphite Powder	kg	21	37
Vinyl Ester Resin	kg	4	7
Steel	kg	5	8

Source: estimated in this study.

Table A.3.6. Stack Material Costs for Mid-Size Vehicles and SUVs

Stack Material	Units	Unit Price* ($/unit)	Material Cost ($)	
			Light-Duty Vehicle	Light-Duty Truck
Membrane (Nafion)	kg	176	142	249
Electrode Catalyst (Pt)	g	28.9	463	811
Gas Diffusion Layer (Carbon Cloth)	kg	30.0	96	168
Bipolar Plate				
Graphite Powder	kg	4.4	94	165
Vinyl Ester Resin	kg	3.7	15	26
Steel	kg	2.9	13	23

* from 2005 PEMFC cost study

Source: estimated in this study.

A.3.5.1.3. Platinum Recycling

In conventional internal combustion engine vehicles, the catalytic converters are recycled to recover the precious metal content (i.e., platinum, palladium, and rhodium), which may be on the order of 5 grams per vehicle depending on the size of the engine. Platinum will continue to be recycled from fuel cell powertrains, but the amount of platinum will be significantly greater, e.g., 0.2 gram per kilowatt net electric or 16 grams for an 80 kW stack. A 2004 study showed that recycling of platinum from fuel cell stacks will be critical to limiting depletion of platinum metal resources.[13] Figure A.3.7 shows how recycling of fuel cell platinum will eventually exceed mined (primary) platinum, given the assumptions of this study. The similarity in slopes for the Total and Recycled curves also shows that transportation fuel cells will become the dominant platinum market. Current markets are approximately split 40/40/20 among auto catalysts/jewelry/industrial, chemical, and electronic markets.

Consequently, the existing transportation infrastructure used to recover catalytic converters will have to expand to handle the increased weight and volume of fuel cell stacks. Additionally, the recycling facilities specifically designed to handle stack materials and volumes will have to be built. The recycling industry for platinum group metal resources is international, with facilities in the end-use countries and in South Africa at the ore refining plants.

The residual value of the platinum group metal resources in fuel cell vehicles will be greater than in current internal combustion engine vehicles due to the higher metal content. The owner of the scrap material may recover on the order of 80% of the market value of the metal content through recycling. The cost of the stack recycling process and recycling value chain will influence the value of the metal content.

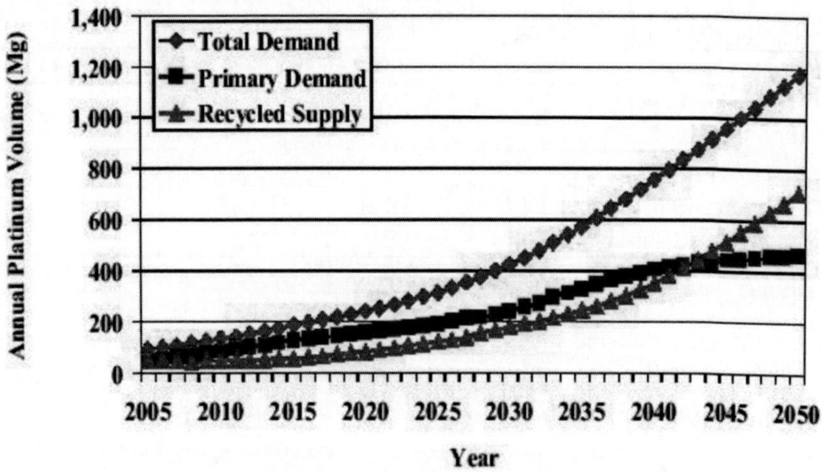

Source: TIAX LLC, "Platinum Availability and Economics for PEMFC Commercialization," report for DOE, December 2003 (Project DE-FC04-01AL67601).

Figure A.3.7. Global FCV Platinum Supply and Demand

A.3.5.2. Hydrogen Storage Sub-system

When the PNGV fuel cell program started, the goal was to use the existing gasoline fuel infrastructure and to convert (reform) gasoline into a hydrogen fuel on-board the vehicle. With the start of the FreedomCAR program, the goal changed to operating the fuel cell on hydrogen stored on the vehicle. Changing from a liquid to gaseous fuel is another significant technology change from conventional internal combustion engine vehicles and has led to a major hydrogen storage technology development program. Compressed hydrogen storage tanks, using carbon fiber designed for pressures of 5,000 and 10,000 psi, are being used in current demonstration vehicles. For lack of an alternative that meets the DOE storage targets, 5,000 psi compressed hydrogen technology has been used.

Table A.3.7 shows the breakout of materials and costs from a 2006 assessment.[14],[15] Carbon fiber is the major weight and cost contributor to the storage system and will continue to be so as the technology evolves. Carbon fiber represents 68% of the carbon fiber layer weight. The 2006 projected cost here ($12/kWh) is significantly higher than the DOE targets of $4/kWh and $2/kWh for 2010 and 2015 respectively.

Table A.3.7. Assessment of Compressed Hydrogen Storage Weight, Volume, and Cost

Basis: 5.6 kg cH2 tank	1 Tank @ 5000PSI		
	Weight kg	Volume liter	System Cost $/kWh
Tank Materials			
Liner	14.40	14.98	0.09
CF Layer	47.54	29.57	7.44
GF Layer	5.31	2.56	0.58
Foam	5.97	10.21	0.22
Tank subtotal	73.22	57.32	8.33
BOStorage (regulators, valving, …) subtotal	13.85	2	3.00
Process Cost ($/kWh)			0.72
CH2	5.89	255.00	0.09
Total	92.96	314.32	12.14
Tank % of Storage System	79%	18%	69%
Estimated			

Table A.3.8. Compressed Hydrogen Storage Assumptions and Calculated Capacity

		Vehicle Type	
	Units	Mid-Size Vehicle	SUV
Energy content of H2	kWh/kg H_2	33.3	
Vehicle range	miles	370	
DOE Target Cost	\$/kWh	4	
Cost of H2	\$/kg	3	
Vehicle Fuel Economy	mpgge[a]	80	58
Tank Capacity[b]	kg	4.6	6.4

[a] miles per gallon of gasoline equivalent; [b] assumes 1 gal gasoline = 1 kg hydrogen. Estimated

The high strength carbon fiber used in this tank specification currently goes into weight- critical aerospace (e.g., airframe components) and energy (e.g., wind turbine blades) applications that require high reliability. Carbon-fiber composite technology and the industry are mature. Major changes in fiber pricing are not expected, and new demand would be met by new plants.

A.3.5.2.1. Hydrogen Storage Material and Cost

A range of 370 miles and fuel economy values (miles per gallon of gasoline equivalent, miles per gge) for the mid-size and SUV fuel cell vehicles were used to calculate the hydrogen storage capacities needed for these vehicles. The required hydrogen storage capacity is an important value because of the cost of hydrogen storage (\$4/kWh or \$133/kg H_2). Table A.3.8 shows the assumptions used in estimating the compressed hydrogen storage tank cost.

A.3.5.3. Hybrid Battery Sub-system

In the time frame of this study, lithium ion battery technology is the most likely candidate for energy storage. In the conventional vehicle, a lead acid battery is used for starting, lighting and ignition. While in a hybrid vehicle, an advanced storage battery technology, e.g., lithium ion or nickel metal hydride, would be used to provide energy for power transients and limited EV operation. For this study, a battery system price of \$375/kWh has been assumed.[16]

A.3.6. OVERALL VEHICLE COST

Figure A.3.8 compares the overall price of the mid-size, conventional and fuel cell versions of the passenger (light-duty vehicle) and the SUV (light-duty truck) vehicles. Also shown are the contributions of the glider, sub-systems, and markups to the overall price. The glider and markups are the two largest cost contributors in both vehicle types. In the future, OEMs may optimize the glider for fuel cell vehicles (as in, for example, the Hy-wire concept vehicle from General Motors) to leverage the characteristics of an all electric powertrain to lower cost, but for now, identical gliders for each vehicle type are assumed. To arrive at a sale price to the customer, the OEM manufacturing cost was marked up by a factor of 1.5 and by a dealer mark-up of 1.16. Dealer mark-ups will vary depending on the vehicle type, but the factor of 1.16 was assumed to represent an industry average across model lines.

Figure A.3.9 shows that the differences between conventional and fuel cell powertrain technology will have limited impact on the overall vehicle price, i.e., on the order of 20% for the light-duty vehicle. This is not surprising if the two technologies need to have comparable prices to be competitive in the market place.

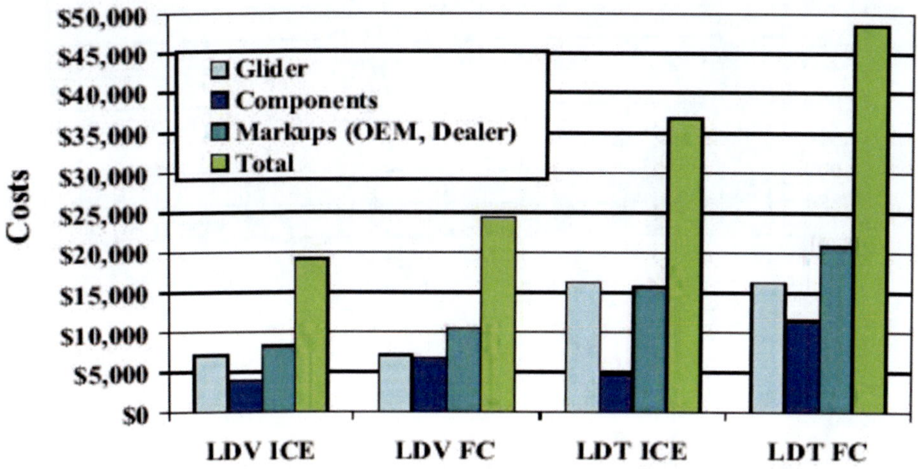

Estimated

Figure A.3.8. Glider, Component and Markup Contributions to the Overall Vehicle Price

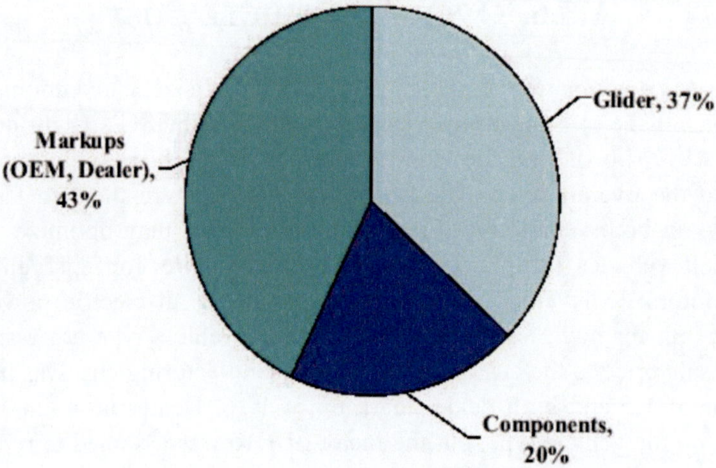

Estimated

Figure A.3.9. Percent Contribution of Glider, Components, and Markups to the Total Vehicle Price

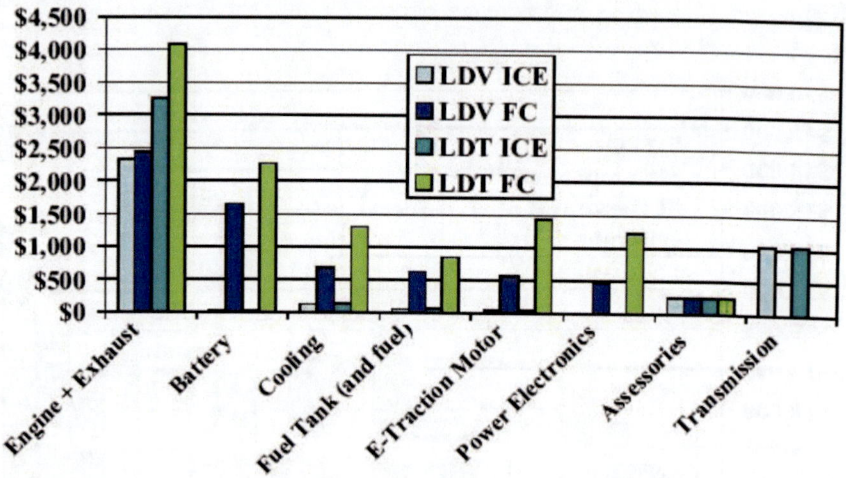

Estimated

Figure A.3.10. Sub-system Cost for the Different Powertrain Configurations

To estimate the impact of a transformation to fuel cell powertrains on manufacturing employment, the changes in vehicle sub-systems have to be considered. For purposes of this discussion, the cost structure is broken down into the following major categories:

Engine and exhaust
Cooling (engine, electronics, battery)
Transmission
Electric traction motor
Accessory power
Battery [Starting, Lighting, and Ignition (SLI); Hybrid]
Fuel Tank (and fuel)

Since the glider is a constant element and a large cost relative to any one of these sub-systems, it is excluded from the following breakdowns. Figure A.3.10 shows the relative contributions of these sub-systems on a cost basis. The categories have been placed in order of descending cost contribution for the mid-size passenger vehicle configuration. The differences in order between the light-duty vehicle and light-duty truck cases arise from the scaling of the electric traction motor relative to the peak power of the combined fuel cell and battery. In the sport utility vehicle (SUV), the motor is sized for extended output from the fuel cell and battery, while in the light-duty vehicle the motor is sized for the fuel cell. This increases the cost of the electric motor and power electronics relative to the other components.

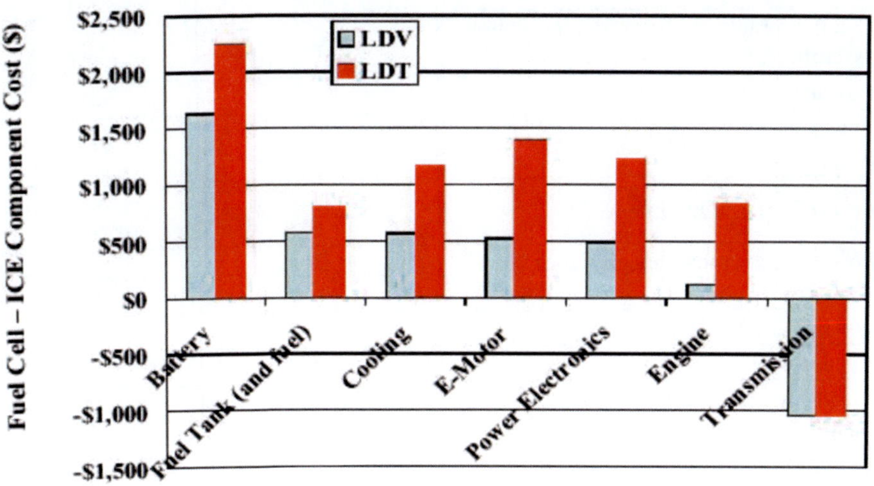

Estimated

Figure A.3.11. Differences in Cost between the Internal Combustion Engine and Fuel Cell Powertrains

Table A.3.9. Detailed Breakdown of Costs, Mid-size Passenger Car

Powertrain and Fuel Configuration	Gasoline Internal Combustion Engine (Baseline)	Hydrogen Fuel Cell Hybrid Electric Vehicle
Glider	$7,148	$7,148
Internal Combustion Engine	$2,077	
Fuel Cell Subsystem, total cost		$2,190
Fuel Cell Subsystem, % materials on cost basis		80%
Fuel Cell Subsystem, materials		$1,752
Fuel Cell Subsystem, design, labor, assembly		$438
Engine Cooling Radiator	$30	$359
Exhaust	$250	$250
Transmission	$1,045	
Electric Traction Motor, total cost	$40	$560
Electric Traction Motor, % materials on cost basis	85%	85%
Electric Traction Motor, materials	$34	$476
Electric Traction Motor, design, labor, assembly	$6	$84
Power Electronics, total cost		$480
Power Electronics, % materials on cost basis		75%
Power Electronics, materials		$360
Power electronics, design, labor, assembly		$120
Power split electronics		
Accessories	$250	$250
Electronics Cooling Radiator		$224
Energy Storage, total cost		$1,019
Energy Storage, % materials on cost basis		85%
Energy Storage, materials		$866
Energy Battery, design, labor, assembly		$153

Powertrain and Fuel Configuration	**Gasoline Internal Combustion Engine (Baseline)**	**Hydrogen Fuel Cell Hybrid Electric Vehicle**
Battery hardware, Acc battery	$20	$633
Energy Storage Cooling Radiator	$90	$103
Main Fuel Storage (Fuel + Tank)	$51	
Carbon fiber tank, total cost		$624
Carbon fiber tank, % materials on cost basis		90%
Carbon fiber tank, materials		$562
Carbon fiber tank, design, labor, assembly		$62
Manufacturing/Assembly Markup ($)	$5,500	$7,045
Dealer Markup ($)	$2,690	$3,445
Overall vehicle price ($)	**$19,191**	**$24,329**

Estimated Vehicle components **meet DOE's goals** for 2015 and beyond. The vehicle system is based on a fuel cell and an energy storage battery. Similar costs would result with a larger fuel cell and no battery.

The fuel cell powertrain has significant increases in cost in most of the major sub-systems as shown in Figure A.3.11. The hybrid fuel cell powertrain has a number of components not present in the conventional vehicle, including a large energy battery, hydrogen storage, electric traction motors, and power electronics. Additionally, the hybrid fuel cell powertrain has increased cooling requirements for the fuel cell, power electronics, and the battery. The fuel cell and the internal combustion engine in the light-duty vehicle are similar in cost because the fuel cell has a lower power rating. When comparably scaled in power, as in the SUV, the fuel cell cost increases relative to the internal combustion engine. The decision was made not to have a transmission in the fuel cell hybrid electric vehicle, but a smaller unit (such as a continuously variable transmission) could be present in an electric powertrain.

Table A.3.9 provides a more detailed breakdown of costs, including the allocation between materials and process costs for a number of the key components, for a mid-size passenger car. Table A.3.10 provides comparable information for the light-duty truck or sport utility vehicle.

The additional cost of the fuel cell vehicles over the conventional vehicles resides primarily in the materials purchased by the companies in the auto sectors. Those material costs contain labor costs, but they are largely upstream in the supply chain, in the chemicals industry, metals sectors, and other sectors selling components to the auto sectors. Consequently the employment impacts will appear in those upstream sectors rather than in the automotive sectors.

Table A.3.10. Detailed Breakdown of Costs, Light-Duty truck / Full-size Sport Utility Vehicle

Powertrain and Fuel Configuration	Gasoline Internal Combustion Engine (Baseline)	Hydrogen Fuel Cell Hybrid Electric Vehicle
Glider	$16,265	$16,265
Internal Combustion Engine	$3,004	
Fuel Cell Subsystem, total cost	$0	$3,833
Fuel Cell Subsystem, % materials on cost basis		85%
Fuel Cell Subsystem, materials		$3,258
Fuel Cell Subsystem, design, labor, assembly		$575
Engine Cooling Radiator	$50	$627
Exhaust	$250	$250
Transmission	$1,045	
Electric Traction Motor, total cost	$40	$1,435
Electric Traction Motor, % materials on cost basis	85%	90%
Electric Traction Motor, materials	$34	$1,292
Electric Traction Motor, design, labor, assembly	$6	$144
Power Electronics, total cost	$0	$1,230
Power Electronics, % materials on cost basis	0%	85%
Power Electronics, materials		$1,046
Power electronics, design, labor, assembly		$185
Power split electronics		
Accessories	$250	$250
Electronics Cooling Radiator		$574
Energy Storage, total cost		$1,610
Energy Storage, % materials on cost basis		85%

Powertrain and Fuel Configuration	**Gasoline Internal Combustion Engine (Baseline)**	**Hydrogen Fuel Cell Hybrid Electric Vehicle**
Energy Battery, design, labor, assembly		$242
Battery hardware, Acc battery	$20	$660
Energy Storage Cooling Radiator	$90	$111
Main Fuel Storage (Fuel + Tank)	$64	
Carbon fiber tank, total cost		$861
Carbon fiber tank, % materials on cost basis		90%
Carbon fiber tank, materials		$775
Carbon fiber tank, design, labor, assembly		$86
Manufacturing/Assembly Markup ($)	$10,539	$13,978
Dealer Markup ($)	$5,154	$6,835
Overall vehicle price ($)	**$36,771**	**$48,520**
Estimated		

End Notes

[1] Global Insight and TIAX LLC, "Light-Duty Future Powertrain Technologies: The Next Generation," Final Report, 2002.

[2] Global Insight and TIAX LLC, "Future Heavy-Duty Powertrain Technologies" The Next Generation," Final Report, December 2004.

[3] The glider is defined as the group of components such as body, chassis, suspension, transmission, and accessories that will not undergo radical change.

[4] TIAX LLC and others, "Comparing the Benefits and Impacts of Hybrid Electric Vehicle Options," report for EPRI, Palo Alto, California, June 2001.

[5] EPRI, "Advanced Batteries for Electric Drive Vehicles," Final Report, Product ID# 1009299, November 2003.

[6] TIAX LLC, *Cost Analysis of PEMFC Systems for Transportation*, NREL/SR-560-39104, December 2005.

[7] Arthur D. Little, Inc., "Guidance for Transportation Technologies: Fuel Choice for Fuel Cell Vehicles, Phase II Final Report," report for DOE, February 2002; available at *http://www1.* eere.energy.gov/hydrogenandfuelcells/pdfs/fuel_choice_fcvs.pdf.

[8] Arthur D. Little, Inc., *ibid.*; Lasher, S., Thijssen, J., Unnasch, S., 2001 Annual Progress Report – Fuels for Advanced CIDI Engines and Fuel Cells, EERE OTT, Washington, DC, November 2001.

[9] Accessed August 2006 at http://www.utcpower.com/fs/com/bin/fs_com_Page/0,11491,047,00.html.

[10] "Well-to-wheels" refers to analysis, also called fuel-cycle analysis, that covers the energy use and emissions of these activities: energy feedstock recovery/production, energy feedstock transportation, fuel production, fuel transportation, distribution, and fueling, and vehicle

operation. With the example of petroleum-based gasoline cycle, the well-to-wheels analysis starts with oil wells in oil fields and ends at energy delivered at vehicle wheels. Source: Agronne National Laboratory

[11] Lasher, S., Unnasch, S., Chan, M., "Energy, Costs, and Transition," 2004 Fuel Cell Seminar, San Antonio, Tex., November 2004.

[12] TIAX LLC, *Cost Analysis of PEMFC Systems for Transportation*, NREL/SR-560-39104, December 2005.

[13] TIAX LLC, "Platinum Availability and Economics for PEMFC Commercialization," report for DOE, December 2003 (Project DE-FC04-01AL67601).

[14] Lasher, S., et al., "Analyses of Hydrogen Storage Materials and On-Board Systems," DOE Annual Hydrogen Merit Review, Crystal City, Virginia May 17, 2006.

[15] Lasher, S., et al., "Comparison of On-board Hydrogen Storage Options," Fuel Cell Seminar, Palm Springs, California, November 2005.

[16] Cost estimates are not quoted for the Li-ion battery cost for hybrid vehicles since they are not commercially available. Cells/batteries have a range of costs from $300/kWh to over a $1000/kWh depending on the performance level, production volumes, and market application for that cell. Valence currently quotes around $700/kWh for batteries that are going into the Segway. To highlight the challenges of most reduction, the estimate of $375/kWh was picked because it is lower than quoted large cell sizes yet higher than long-term cost targets. The cost used also includes the components required to safely operate a large battery.

APPENDIX 4. INPUT-OUTPUT MODELING OF THE EMPLOYMENT IMPACTS OF THE HYDROGEN ECONOMY

This appendix (Appendix 4) describes the input-output modeling system used to estimate the employment impacts of the hydrogen economy. The hydrogen scenarios and technology cost estimates described in Appendices 1 through 3 formed the basic inputs and assumptions used to initiate the model simulation.

Appendix 4 is divided into seven sections. The first section provides an overview of input-output analysis and how it is used to estimate the employment impacts of the expansion of hydrogen markets. The second section provides background on the input-output model that formed the core of the modeling process, IMPLAN. The third section describes the overall modeling process and includes a step-by-step overview of the methodology and key data inputs. The fourth section describes the methodology employed to forecast the model to the 2020, 2035 and 2050 forecast years. The fifth section provides an overview of each of the input and construction vectors that were created in order to change the economic relationships and spending patterns to reflect the hydrogen economy. The sixth section provides details on how the input vectors (production functions) were altered or created from the technology cost estimates to reflect the production of hydrogen vehicles and hydrogen fuels. The seventh section describes how the estimates of construction costs and construction costs avoided were developed based on the technology cost estimates and other sources.

A.4.1. THE USE OF INPUT-OUTPUT ANALYSIS

Input-output analysis models the inter-industry dependencies in an economy. The model begins with a table that quantifies the value of purchases each industry makes from each other industry, as well as the purchases by households, government, and foreigners from each industry. Industries are both customers and suppliers of one another, and they hire labor services from households in the form of employment. Households, as well as governments, industries making investments, and foreign buyers, also purchase the products of each industry, in sales known as final demand. The inter-industry dependencies are studied by changing final demands. A change in the final demand facing a particular industry causes that industry to change its purchases of inputs from each of its intermediate suppliers, and its purchases of labor services from households. Those industries, in turn, must increase their purchases from their own suppliers, and so on through multiple rounds of spending that ripple through the economy. Changes in household income caused by increases or decreases in employment can also affect subsequent rounds of final demand purchases.

The core of an input-output model is the transactions matrix—the table of each industry's purchases from each other industry. Reading down each column of the table, an industry's purchases from each other industry are reported. Reading across each row, an industry's sales to each other industry are reported. Presently no industry produces enough of the hydrogen vehicles and related products, fuel, and infrastructure for even the most recent input-output table available (2002) to describe the inter-industry purchase pattern of a hydrogen economy. To modify the purchase table to account for the hydrogen economy, the cost data reported in Appendices 2 and 3 were used to modify the purchase vectors of three key industries—motor vehicle components manufacturing, motor vehicle assembly, and hydrogen production.

The employment impacts of the transformation of the economy from gasoline to hydrogen were estimated by running the IMPLAN model to simulate the economy's employment structure at 2020, 2035, and 2050 under the three scenarios described in section 2—the base case that is a predominately gasoline economy, the HFI Scenario of a rapid transformation to a hydrogen economy, and the Less Aggressive Scenario. At each date, the difference in employment between one of the hydrogen scenarios and the base case is the impact of the hydrogen economy at that date.

A.4.2. DESCRIPTION OF THE IMPLAN MODEL

IMPLAN (IMpact Analysis for PLANning) is an input-output (I-O) model. The most recent (2003) version of this model was selected for this project for its detailed sectoring plan (509 sectors), its ability to trace impacts through the economy, and its ability to model impacts at the regional level. The following paragraphs provide an overview of the model. The majority of this discussion was derived from IMPLAN model documentation.[1]

IMPLAN was originally developed by the USDA Forest Service in cooperation with the Federal Emergency Management Agency and the USDI Bureau of Land Management to assist the Forest Service in land and resource management planning. The IMPLAN system has been in use since 1979 and has evolved from a main-frame, non-interactive application that ran in "batch" mode to a menu-driven microcomputer program that is completely interactive.

The Minnesota IMPLAN Group began work on IMPLAN databases in 1987 at the University of Minnesota. In 1993, Minnesota IMPLAN Group, Inc. (MIG) was formed to privatize the development of IMPLAN data and software. Version 1 of the Windows software was developed by MIG and released in June of 1996. Version 2 was released in May 1999.

IMPLAN Professional introduces flexibility in the methods and assumptions used to generate social accounts and I/O multipliers and takes full advantage of the Windows environment not found in other systems.

There are two components to the IMPLAN system, the software and the database. The software performs the necessary calculations, using the data for either a national or a regional study, to create the models. It also provides an interface for the user to change the studied region's economic description, create impact scenarios and introduce changes to the local model. The databases provide all the information needed to create regional IMPLAN models.

The IMPLAN database, created by MIG, Inc., consists of two major parts: 1) national-level technology matrices; and 2) estimates of regional data for institutional demand and transfers, value-added industry output and employment for each county in the United States as well as state and national totals. The IMPLAN data and accounts closely follow the accounting conventions used in the "Input-Output Study of the U.S. Economy," a study by the Bureau of Economic Analysis (1980), and the rectangular format recommended by the United Nations. Comprehensive and detailed data coverage of the entire United States, by county, and the ability to incorporate user-supplied data at each stage of the model building process, provide a high degree of flexibility both in terms of geographic coverage and model formulation.

A.4.3. DESCRIPTION OF THE INPUT/OUTPUT MODELING PROCESS

As hydrogen-based transportation is introduced, the structure of the economy will change from a baseline economy to the hydrogen economy. At the same time, demand for purchases and investments in gasoline technologies will give way to demand for purchases and investments in hydrogen technologies.

This section provides a step-by-step overview of the modeling process, showing the mathematical steps involved in producing estimates of employment in the three scenarios. At the end of each step, the sources of the required data inputs are specified. Sub-sections A.4.3.1 through A.4.6.1 provide additional information on portions of the process that are too detailed to be included in this section.

A.4.3.1. Overview of the Process

The change in the structure of the economy is modeled by changing the coefficients for the directly affected sectors of the economy in the IMPLAN national absorption coefficient (use) matrix. For example, Table A.4.1 shows an abbreviated list of the input coefficients to Sector 148 (Industrial Gas Producers). The uses in column A include both domestic and imported goods, which are reported in more detail in columns D through G. Column B contains the "use coefficient vector" for industrial gas production, also called the production function.

A production function represents the relationship between the inputs to a productive process and the outputs that result. Traditional production functions relate inputs of capital and labor to the production of particular products according to some mathematical formulation. Input-output (I/O) production functions relate commodity inputs and labor to industry outputs in a linear manner. Any type of production function is specific to the technology in use. When that technology changes, the characteristics of the production function change as well. The production function shows the current schedule of inputs required for $1 of output from this sector.

The IMPLAN national use matrix has 509 columns, with production functions for each producing sector in the model. Since all sectors interact with all other sectors, any change in the use matrix causes a structural change in output and employment patterns throughout the economy. The IMPLAN model has only a national use matrix. The national use coefficient matrix is used in each regional model. In the regional models, the inputs may be produced locally, or imported from domestic sources outside the region or from foreign sources.

Once a new structural matrix is developed that reflects the hydrogen technologies, a revised set of final demands for purchases and investments can be introduced. When the new structural model simulation is initiated by these demands, the model will specify the new employment required throughout the economy. These results can be compared with the employment generated in the base-case economy to estimate the employment effects of the hydrogen economy.

A.4.3.2. Baseline Model (Maintain *status quo* structural matrix through 2050)

The 2003 IMPLAN national and regional I/O models define the structure of the *baseline economy*, including current inputs, outputs, and employment for 509 sectors. The structural coefficients in the baseline model provide a set of linear equations that are solved to determine the amount of output from each sector required to supply demand from end-users, including households, state and local government, the Federal government, and exports. In this project, the household sector is the largest source of final demand for hydrogen vehicles and hydrogen. Other sources of demand that are modeled include changes in gross private fixed investment (GPFI) that result from construction of hydrogen production facilities, hydrogen service stations, auto parts plants, coal mines, and the avoided costs relating to the reduced demand for the construction of refineries and gasoline service stations.

Table A.4.1. Use Matrix for Industrial Gas Manufacturing, IMPLAN Sector 148 [in millions of 2005 dollars] (Cutoff = $5 Million of Total Use)

		[A]	[B]	[C]	[D]	[E]	[F]	[G]
Commodity	Commodity Name	Total Use	% of Output	% of Inputs	Local Use	% of Local	Imports	% from Imports
19	Oil and gas extraction	33.22	0.55	2.07	17.71	1.13	15.51	46.69
30	Power generation and supply	509.82	8.42	31.83	508.20	32.58	1.62	0.32
31	Natural gas distribution	72.22	1.19	4.51	72.15	4.63	0.07	0.10
43	Maintenance and repair of nonresidential building	28.29	0.47	1.77	28.29	1.81	0.00	0.00
142	Petroleum refineries	24.79	0.41	1.55	23.75	1.52	1.04	4.19
148	Industrial gas manufacturing	14.03	0.23	0.88	13.67	0.88	0.36	2.56
243	Machine shops	8.44	0.14	0.53	8.43	0.54	0.01	0.14
248	Metal valve manufacturing	11.26	0.19	0.70	7.93	0.51	3.33	29.58
311	Semiconductors and related device manufacturing	13.63	0.22	0.85	9.99	0.64	3.64	26.69
312	All other electronic component manufacturing	12.66	0.22	0.79	7.75	0.50	4.91	38.80
390	Wholesale trade	75.88	1.25	4.74	75.88	4.86	0.00	0.00
391	Air transportation	6.20	0.10	0.39	5.24	0.34	0.96	15.49
394	Truck transportation	5.80	0.10	0.36	5.73	0.37	0.07	1.22
397	Scenic and sightseeing transportation and sup	9.43	0.15	0.59	9.38	0.60	0.05	0.58
400	Warehousing and storage	33.33	0.55	2.08	33.33	2.14	0.00	0.00

Table A.4.1. (Continued)

		[A]	[B]	[C]	[D]	[E]	[F]	[G]
Commodity	**Commodity Name**	**Total Use**	**% of Output**	**% of Inputs**	**Local Use**	**% of Local**	**Imports**	**% from Imports**
422	Telecommunications	5.58	0.09	0.35	5.58	0.36	0.00	0.01
424	Data processing services	13.78	0.23	0.86	13.77	0.88	0.00	0.03
425	Nondepository credit intermediation and related institutions	12.04	0.20	0.75	12.03	0.77	0.00	0.03
426	Securities; commodity contracts; investments	5.62	0.09	0.35	5.62	0.36	0.00	0.00
427	Insurance carriers	9.00	0.15	0.56	8.82	0.57	0.19	2.06
430	Monetary authorities and depository credit in	21.81	0.36	1.36	21.77	1.40	0.02	0.10
431	Real estate	5.84	0.10	0.36	5.84	0.37	0.00	0.00
436	Lessors of nonfinancial intangible assets	150.28	2.48	9.38	150.25	9.63	0.03	0.02
437	Legal services	24.41	0.40	1.52	24.32	1.56	0.10	0.42
438	Accounting and bookkeeping services	7.60	0.13	0.47	7.55	0.48	0.04	0.60
439	Architectural and engineering services	57.81	0.95	3.61	57.44	3.68	0.37	0.64
444	Management consulting services	6.53	0.11	0.41	6.42	0.41	0.12	1.85
446	Scientific research and development services	28.77	0.47	1.80	28.54	1.83	0.24	0.81
450	All other miscellaneous professional and tech	14.81	0.24	0.92	14.81	0.95	0.00	0.00

Table A.4.1. (Continued)

		[A]	[B]	[C]	[D]	[E]	[F]	[G]
Commodity	**Commodity Name**	**Total Use**	**% of Output**	**% of Inputs**	**Local Use**	**% of Local**	**Imports**	**% from Imports**
451	Management of companies and enterprises	204.23	3.37	12.75	204.23	13.09	0.00	0.00
481	Food services and drinking places	7.48	0.12	0.47	7.48	0.48	0.00	0.00
483	Automotive repair and maintenance; except car	5.90	0.10	0.37	5.90	0.38	0.00	0.00
484	Electronic equipment repair and maintenance	10.36	0.17	0.65	10.36	0.66	0.00	0.00
485	Commercial machinery repair and maintenance	19.62	0.32	1.22	19.62	1.26	0.00	0.00
499	Other State and local government enterprises	8.27	0.14	0.52	8.27	0.53	0.00	0.01
	Total Inputs (All Commodities)	1,603.73	26.45	100.00	1,560.02	100.00	41.83	2.61
	Value Added	4,459.08	73.55					
	Industry Output	6,062.81	100.00					

Source: Minnesota IMPLAN Group, Inc., *IMPLAN Professional, Version 2.0, Social Accounting &Impact Analysis Software*,2002 national input-output table.

Table A.4.2. Input Data Requirements

Data Item	Source
IMPLAN National and selected regional I/O models, 2002 data	Purchased from MIG
Population and Final demand growth rates (2006 to 2020, 2020 to 2035, 2035 to 2050)	Abstracted from DOE Posture Plan and REMI Model (see section A.3.3)
Transportation demand growth scenario for motor gasoline	Abstracted from DOE Posture Plan
Baseline Employment Model (includes productivity growth through 2050)	

The baseline model is used to project baseline output and employment through 2050, thereby reflecting growth in real final demand and increases in labor productivity beyond current levels, while holding fixed the structural coefficients in the model.

Due to the linear nature of the I/O model, the baseline use coefficient matrix may be solved for three arrays of *baseline multipliers* $M_i(0)$, i=1, 2, 3. (The argument of 0 indicates that these are the baseline matrices for the baseline case without hydrogen.) The baseline multipliers show the direct, indirect, and induced effects on each sector of the economy that result from a specified change in the final demand vector. The direct impacts are impacts on directly affected sectors. Indirect impacts include effects on the suppliers to the directly affected industries, and upon their upstream suppliers in turn, and so on, in continuing rounds of spending. Induced impacts arise as a result of the additional (or reduced) purchasing power from households receiving the increased (or reduced) wages that result from the direct and indirect changes in output. Mathematically, the direct, indirect, and induced impacts are computed using matrix multiplication:

$$\Delta \underline{x}_i(t, 0) = M_i(0)\, \Delta \underline{y}(t, 0),$$

where $\Delta \underline{y}(t, 0)$ is the *baseline final demand change* from current levels by year t, $M_i(0)$ is the baseline multiplier matrix of type i, and $\Delta \underline{x}_i(t, 0)$ is the resulting *baseline output change* for each sector in year t from current levels. While the primary interest in this study is with the changes in sector outcomes, it is important that expected levels of activity for each sector are reasonable. Therefore, assumptions are required regarding the expected changes in exogenous components of aggregate demand and supply through 2050. The sources and methods of these adjustments to the model are covered in detail below in Section A.4.3.

Table A.4.2 below describes the input data requirements for this portion of the model.

A.4.3.3. Baseline Employment Model (Includes Productivity Improvement through 2050)

Employment is assumed proportional to output in the traditional I/O model. Direct application of these employment coefficients provides the basis for initial estimates of the direct, indirect, and induced changes in employment, assuming fixed labor productivity. The IMPLAN model includes employment-to-output ratios for each sector. These ratios are placed on the diagonal of a matrix with zeroes off the diagonal. The product of this *baseline employment coefficient matrix*, E(0), and the respective output changes, $E(0)\ \Delta\underline{x}_i(t, 0)$, yields the baseline direct, indirect, and induced employment estimates in year t for fixed labor productivity. The initial estimates of baseline employment were corrected for productivity gains by adjusting the employment coefficient matrix in future years:

$$\Delta\underline{e}_i(t, 0) = E(t)\ \Delta\underline{x}_i\ (t, 0).$$

Here $E(t) = E(0)\ (1\text{-}p)^t$ is the *productivity-adjusted employment coefficient matrix,* where p is the annual rate of growth in labor productivity. Different rates of productivity change were applied to different groups of sectors of the economy, using data described in Section A.4.3, below. The results of the baseline modeling effort include baseline output and employment projections by sector for the *status quo* economy in each selected year.

Table A.4.3 below describes the input data requirements for this portion of the model.

A.4.3.4. Hydrogen Technology Structural Models

Analysis of the alternative hydrogen production technologies required creation of a new national use matrix that combines the technologies shown in Table A.4.4. The production function for the directly affected sector (148, Industrial Gases) in the baseline use matrix was changed to reflect the new hydrogen technologies. The national use coefficient matrix for a given technology will be applied in all years. National use coefficients reflect production, distribution, and delivery costs. Where there is more than one possible choice of distribution and delivery channel, fixed proportions are assumed, e. g., 50% truck and 50% pipeline.

The use matrix in the technology model differs from the baseline use matrix in several directly-affected sectors of the economy. The most significant changes are in hydrogen production and delivery, light-duty vehicle production, and auto parts production. Although there are 509 sectors in the IMPLAN model, only a small number of sectors are directly affected. Hydrogen gas producers are classified in IMPLAN Sector 148 (Industrial Gas Producers). They presently account for only a small portion of this small sector. This sector was renamed Hydrogen Fuel Producers, and required major changes in its use vector to reflect the inputs for the combination of hydrogen production technologies in each year and scenario.

Table A.4.3. Input Data Requirements

Data Item	Source
IMPLAN Employment data, National and selected regional, 2003 data	Purchased from MIG
Labor productivity improvement rates by sector group for 2005-2020, 2020-2035, 2035-2050	REMI Model (see section A.4.3)

Table A.4.4. Hydrogen Production Technologies in the New National Use Matrix, with Years of Appearance

Production Technology	Scenario Years
Steam methane reforming (SMR), 1500 kg/day distributed stations	2020, 2035, 2050
Centralized SMR, 380K kg/day (with carbon sequestration)	2035
Coal gasification, 307 kg/day (with carbon sequestration)	2035, 2050
Centralized biomass, 155K kg/day	2035, 2050
Nuclear electrolysis	2035, 2050
Nuclear thermochemical	2035, 2050
Wind	2035, 2050
Electrolysis, 1500 kg/day distributed stations	2020

The use vectors for the auto sector and the auto parts sector were also changed from the baseline. In each year and scenario, the vectors for hydrogen vehicle and auto parts were weighted with the vector for the existing auto sectors in proportion to vehicle production. The delivery of hydrogen to end-users was assigned to the current transportation, wholesale, and service station sectors. Additional costs incurred by these sectors to satisfy hydrogen technological requirements are included within the new Hydrogen Fuel Producers sector.

Additional details on the changes in the vectors by sector are provided below in Sections A.4.4 and A.4.5.

The following steps were required to determine the multiplier sets for the national models in each year and scenario:

1. Develop a new vector for each type of hydrogen production, for each type of vehicle production (auto, light truck/SUV), and for each type of vehicle parts production (auto, light truck/SUV).
2. For each sector (hydrogen production, vehicle production, and parts production) create a combined vector by weighting production of the various technologies based on penetration rates for that technology in that year and scenario)
3. Input the new technologies (six different year and scenario combinations) into the IMPLAN model:

Year and Scenario → Model (Year, Scenario) → Multiplier set M_i(Year, Scenario)

Revise baseline use vector for IMPLAN sector 148 (Hydrogen producers) Revise baseline use vector for IMPLAN sector 344 (Auto production)

Revise baseline use vector for IMPLAN sector 350 (Auto parts)

Changing the use matrix generates a new model of the economy and a new set of multipliers for each technology. The new multipliers reflect the new structure of the economy due to hydrogen penetration in the transportation sector. Multiplier sets for each technology were calculated using the IMPLAN national and regional models and exported to ASCII text files. The large number of matrix multiplications required to determine productivity-adjusted employment impacts were done outside the IMPLAN package.

Table A.4.5 below describes the input data requirements for this portion of the model.

Table A.4.5. Input Data Requirements

Data Item	Source
Penetration Rates for Hydrogen Technologies	DOE Posture Plan
Use coefficient vectors for hydrogen production	See Section A.4.5
Use coefficient vectors for auto and auto parts production	See Section A.4.5

A.4.3.5. Hydrogen Technology Demand

The multipliers for each year and scenario were applied to a revised set of final demand change vectors that reflect the penetration of hydrogen transportation in each year and corresponding scenario. Separate final demand vectors were prepared for household transportation demand and construction costs for each year and scenario. For each household final demand schedule, a separate schedule of construction of hydrogen production and delivery facilities was scaled to accommodate the level of hydrogen-based transportation demand. Construction cost savings for the additional petroleum-based transportation facilities that were built in the baseline economy, but are not required in the hydrogen-based economy, were factored in proportion to the petroleum-based transportation demand avoided.

The primary difference in the hydrogen economy final demand vectors from the baseline is less transportation fuel demand from the petroleum refining sector, replaced with hydrogen from the newly named hydrogen production industry. In IMPLAN, levels of final demand are specified by sector in producer value, with additional retail, wholesale, and transportation margins assigned to the appropriate sectors. Table A.4.6. Input Data Requirements

Table A.4.5. Input Data Requirements

Data Item	Source
Construction cost estimates (new and avoided)	See Section A.4.6
Hydrogen demand schedule, 2020, 2035, 2050	Appendix 1
Gasoline demand schedule, 2020, 2035, 2050	Appendix 1
Auto and light truck/SUV demand schedules 2020, 2035, 2050	Appendix 1

Consumers will continue to purchase autos from the auto sector with current IMPLAN retail margins by region, but the auto sector will use a different mix of inputs to make vehicles. Similarly, consumers continue to pay margins to the service station sector, wholesale, and transportation sectors for purchased fuel.

A.4.3.6. Output and Employment Impacts

Output changes from baseline levels in year t and scenario A, Δx_i (t, A), are determined by multiplication of the appropriate multiplier matrix by the final demand change vector for this technology:

$$\Delta \underline{x}_i (t, A) = M_i(A)\, \Delta \underline{y}(t, A).$$

The direct, indirect, and induced employment impacts due to adoption of year t under scenario A are measured by the net change in employment levels by sector from baseline levels:

$$\Delta \underline{e}_i(t, A) = E(t)\, [\Delta \underline{x}_i (t, A) - \Delta \underline{x}_i (t, 0)].$$

Labor productivity improvements introduced in the baseline also are applied to the employment estimates for the new hydrogen economy when computing the net change in employment.

A.4.4. Forecasting the Base Case Model

The IMPLAN model is a basic static equilibrium model calibrated to 2003 estimates of the national components of aggregate supply and demand by sector with an employment link attached. This model is most useful for analyzing exogenous changes in the current economy. When applied to exogenous changes over a lengthy period of time, forecasts of change over time must be made as inputs into IMPLAN in order to derive the base case scenario from which the future impacts of the exogenous changes in later years can be evaluated. While the primary interest in this study is with the changes in sector outcomes, it is important that expected levels of activity for each sector are reasonable, so that the changes themselves are as accurate as possible and can be put in context. Thus, there is a need to make some assumptions regarding the expected changes in exogenous components of aggregate demand and supply through 2050 for application to the IMPLAN model.

To accomplish this, one could simply calibrate the model to the national growth rates explicit in DOE analyses, particularly growth in real GDP. The problem with this approach is that all of the transaction matrix coefficients with the IMPLAN model will remain the same so that all sectors will grow evenly at the same rate as aggregate national demand. The problem with this approach is that such a growth pattern is not likely to occur. Over the past 20 years there have been significant shifts in the market shares of sectors. For example, manufacturing's share of total U.S. output has been declining while the service sector's share has been rising. Thus, in order to make IMPLAN's 2050 results as realistic as possible, changes in the very structure of the U.S. economy must be imbedded into the IMPLAN transaction matrix.

In addressing this problem, the first question is what existing sources might be relied upon to forecast growth by sector over the next 45 years. Most governmental agencies now restrict their growth forecasts by either a high-level industrial classification or to just a few years. Federal agencies such as the Census Bureau, the Bureau of Labor Statistics (BLS) or the Bureau of Economic Analysis (BVEA) do not provide 50-year forecasts by detailed sector. However, the study team had access to the REMI model's national baseline forecast for 2006.[2] This model, produced by Regional Economic Models, Inc., is the most widely cited of the privately produced I-O hybrid models in the country, and its results and methodology have been published in peer-reviewed academic journals. Furthermore, the REMI model relies upon BLS statistics, trends, and short-run forecasts. As a start, the REMI model's forecasts for employment, aggregate demands, and supplies by sector were used as initial forecasts for the IMPLAN model. Aggregations were required since the REMI model has only 72 sectors and IMPLAN has 509. Within-group sectors were treated the same in terms of growth potential.

The use of REMI forecasts provided an excellent beginning in the construction of an IMPLAN base case (without hydrogen) simulation. However, REMI too has some significant weaknesses. Whereas IMPLAN fixes the market shares of all sectors to their 2003 levels, REMI essentially fixes past changes in these market shares into the future. Thus, a sector experiencing significant, above-average growth in the past 10 years will show that same tendency for the next 45 years. Obviously, this will not always be the case. For example, pertinent to this study are the REMI outcomes regarding energy. Essentially, the growth potential of oil and gas extraction, the manufacturing of oil and gas exploration equipment, and exploration for oil and gas were forecast to be modest because their recent history (1993 to 2003) had been modest. Recent events suggest otherwise. Thus, in some cases additional information was required in order to fine tune the REMI output results and inputs into IMPLAN.

The initial demand vectors used from REMI satisfactorily produced 2050 forecasts for all 509 IMPLAN sectors that guided reasonable estimates of economy-wide changes due to the transformation of the U.S. economy towards hydrogen. Indeed, it should be recognized that forecasts this far in the future can vary substantially between analysts and have even varied significantly at the more aggregate level in forecasts by DOE over the years.

The point of the exercise was not so much to provide a precise and detailed forecast of the economic future of the nation, but to create a broad-brush projection of the national economy in 2050, the outlines of which were broadly consistent with most knowledgeable analysts' expectations of an ever-evolving economy. More attention was given to key sectors directly related to the conventional energy sectors or the new hydrogen sector, to make sure that these sectors were growing in accordance with consensus beliefs. All sectors were also adjusted to guarantee that the macro statistics on aggregate demand, supply, and trade were consistent with DOE expectations and the expectations of expert professionals utilized as consultants on this project.

In some cases, the IMPLAN results produced growth rates that were unrealistically high or low. This pattern could be easily identified as having been driven by the fact that these sectors had recently experienced extraordinarily high or low growth rates which were not likely to continue. In this case, longer term growth rates as implied by BLS statistics which went back to the 1980s and 1990s were examined to see what longer term trends had been. In some cases, such as the extraordinarily high gains in labor productivity in the refining industry, the study relied upon industry experts for guidance. The universal consensus regarding refining was that while further productivity gains were expected, the rate of gain in productivity of the 1990s and early part of this decade (2000 to 2010) should not be expected to continue for the next 45 years.

The end result was an IMPLAN 2050 baseline output that showed real GDP growing at 2.5% per year on average, employment growth at 1.6% per year, and both exports and imports growing, but in such a way that the U.S. trade deficit falls gradually over time as a percent of real GDP. All of these assumptions are consistent with DOE assumptions.

The REMI data include forecasts of output and employment for the years 2005, 2020, 2035, and 2050, and forecasts of final demand, import, and exports for the same years. Average annual growth rates by sector were computed for the periods 2005-2020, 2020-2035 and 2035-2050. The growth rates were applied to the IMPLAN baseline matrices for the periods 2003-2005, 2005-2020, 2020-2035, and 2035-2050. Growth between 2003 and 2005 was imputed using the REMI growth rate in the first available time period.

The final demand growth rates were applied to all components of final demand, except imports to final demand and exports. (These components include: Households, Federal and state/local government, Investment and Inventory Change.) Separately identified REMI growth rates were used for imports going to final demand and exports.

Output growth rates were applied to the 2003 IMPLAN output. The intermediate uses and the value added components for each industry were then scaled using fixed 2003 IMPLAN use coefficients to accommodate the new level of output in each year. Employment in each year was computed using IMPLAN 2003 output/employment ratios for each sector. Employment in each sector then was adjusted in each year using the growth in labor productivity implied by the REMI forecasts of output and employment in the corresponding time period. The forecast was compared with the HFI Scenario output and population forecasts.

After fixing growth for output, final demand, imports to final demand, and exports, the only part of the I/O model without an exogenous forecast is that of the intermediate inputs. These include both imports used as intermediate inputs (crude oil, for example) and domestically produced inputs to production. The inputs to all 509 industries were calculated using the fixed IMPLAN 2003 production functions. Currently, the model is not in balance, since the production functions for all industries would have to be changed in each year to correct the imbalances. For example, in the initial runs, imbalances, summed over the 4 forecast years (2005, 2020, 2035 and 2050), totaled about $8 trillion or 4 percent of total demand in these years. Imbalances tend to grow in the later years of the forecast period.

A.4.5. Industry Distribution of Vehicle Costs

An input-output model requires the final consumer cost of vehicles to be distributed to all of the industries contributing to the final appearance of the finished vehicle in the showroom. Table A.4.7 reports the distribution of production and operating costs and the allocation of user costs to taxes, for the base case and the HFI Scenario in 2050. In the hydrogen economy, it is assumed that there are no gasoline vehicles sold in 2050 (there still are gasoline vehicles around). The base case assumes that the hydrogen demand is extremely small, essentially zero. Figures A.4.1 through A.4.2.6 show comparable distributions of costs for both hydrogen scenarios for each of the three years. Figure A.4.6 shows the distributions of Table A.4.7 graphically.

Table A.4.7. Industry Distribution of Total Annual Vehicle Costs for the HFI Scenario, 2050 [in millions of 2005 dollars]

Industry, with IMPLAN sector numbers	Base Case, Gasoline Vehicles	HFI Scenario	
		Hydrogen Vehicles	Gasoline Vehicles
Auto Manufacturing (350)	$584,464	$751,578	$0
Auto Dealers (401)	$89,283	$117,497	$0
Rail (392)	$2,394	$2,394	$0
Water (393)	$7,161	$1,197	$228
Truck (394)	$36,794	$14,058	$1,313
Pipeline (396)	$5,964	$8,823	$228
Service Stations (407)	$20,765	$13,703	$793
Gasoline Production (142)	$415,629		$15,868
Hydrogen Production (148)		$172,389	
Fuel Road Tax	$61,448	$22,206	$2,346
Hydrogen Parity Tax[a]	$0	$36,896	$0
Vehicle Sales Tax	$40,425	$52,145	$0
Fuel Sales Tax	$36,066	$13,034	$1,377

[a] Hypothetical tax increment applied to hydrogen vehicles to maintain tax neutrality with gasoline vehicles.

Estimated

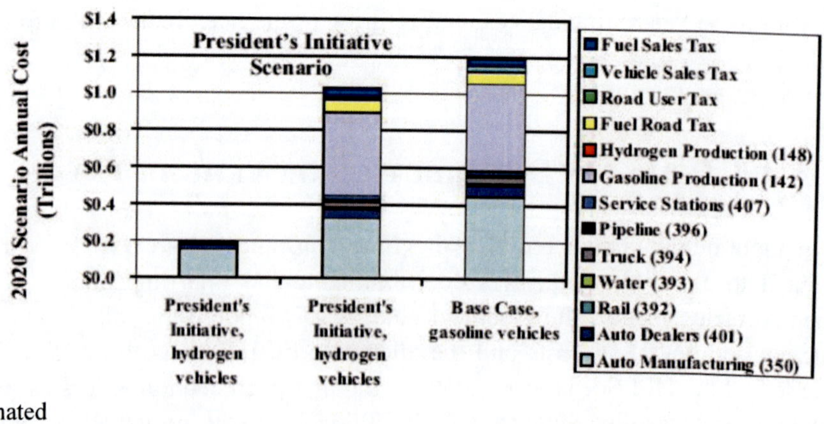

Estimated

Figure A.4.1. Distribution of Total Vehicle Costs in the HFI Scenario, 2020

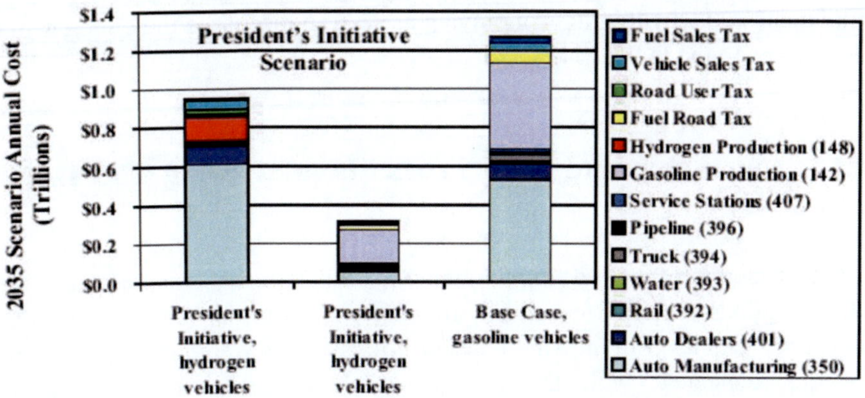

Estimated

Figure A.4.2. Distribution of Total Vehicle Costs in the HFI Scenario, 2035

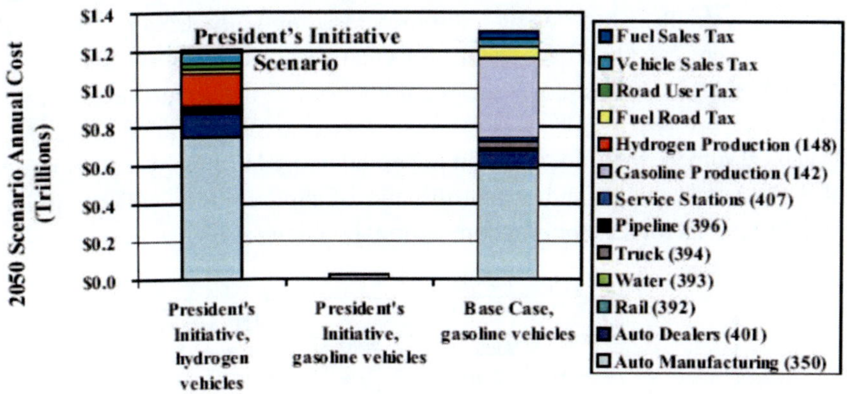

Estimated

Figure A.4.3. Distribution of Total U.S. Vehicle Costs in the HFI Scenario, 2050

A.4.6. Overview of Adjusted Input and Construction Vectors

Estimates of revised production functions and final demand vectors were required to develop and initiate the simulation of the IMPLAN model to estimate employment in the hydrogen economy. This section discusses the changes in each of the sectors in the IMPLAN model.

A.4.6.1. Automobile and Light Truck Manufacturing (IMPLAN Sector #344)

The production function (input vector) for this industry was altered to reflect the bill of goods required to build hydrogen vehicles rather than current internal combustion vehicles. Engineering data for both systems were analyzed to subtract the costs associated with the internal combustion power train and other components and add in the costs associated with the power train and other components. Data for a typical passenger car and light duty trucks/SUVs were developed and weighted with the original production function for internal combustion vehicles in developing these estimates. The weights for internal combustion vehicles, hydrogen automobiles, and hydrogen light duty trucks/SUVs were calculated separately for each year and scenario. Section A.4.4 provides detailed information on the manipulation of the input vectors for this sector.

It was assumed that no additional capital expenditures were required to assemble the new vehicles and/or that the mix of capital goods will be similar to the capital required for assembling conventional vehicles.

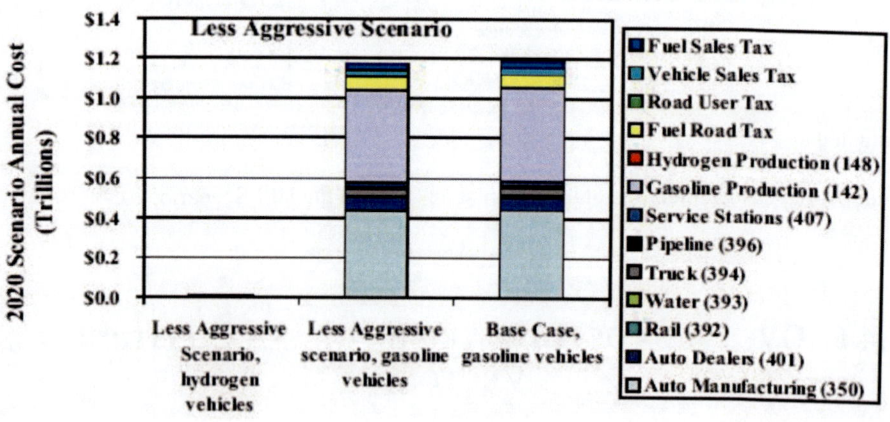

Estimated

Figure A.4.4. Distribution of Total Vehicle Costs in the Less Aggressive Scenario, 2020

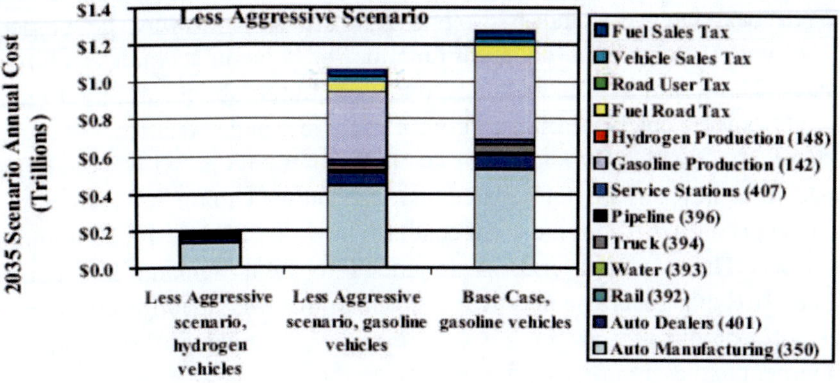

Estimated

Figure A.4.5. Distribution of Total Vehicle Costs in the Less Aggressive Scenario, 2035

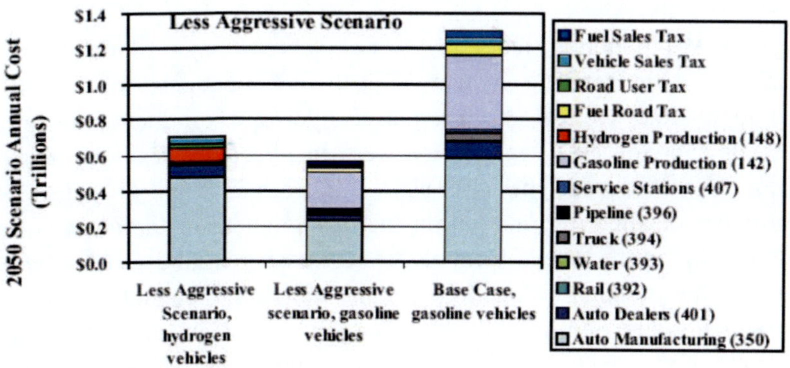

Estimated

Figure A.4.6. Distribution of Total Vehicle Costs in the Less Aggressive Scenario, 2050

A.4.6.2. Motor Vehicle Parts Manufacturing (IMPLAN Sector #350)

The production function (input vector) for this industry was altered to reflect the bill of goods required to build hydrogen vehicle parts rather than current internal combustion (IC) vehicle parts. Engineering data for both systems were analyzed to subtract the costs associated with producing components for IC vehicles and to add in the costs associated with the hydrogen vehicle components. As was the case for Automobile and Light Truck Manufacturing, data for a typical hydrogen passenger car and hydrogen light truck/SUV were developed and weighted with the original production function for IC vehicles in developing these estimates. The weights for IC vehicles, hydrogen automobiles, and hydrogen light trucks/SUVs were calculated separately for each year and scenario.

It was assumed that additional capital expenditures will be required to produce the new hydrogen related parts. These capital investments were developed primarily to add in capital costs for new facilities designed to build fuel cell stacks. These capital costs were added to final demand and consist of estimates of both general construction costs as well as specific costs for large pieces of equipment. Section A.4.7 provides detailed information on the development and assignment of these costs.

A.4.6.3. Industrial Gas Manufacturing (IMPLAN Sector #148)

Eight new production functions (input vectors) were developed for the industrial gas manufacturing industry. Each vector reflects the bill of goods required to operate each of the alternative hydrogen production technologies. Engineering data for each technology were analyzed to develop the costs of production with overhead type costs estimated from similar I-O sectors. Section A.4.5 provides detailed information on the development of the input vectors for this sector. Data on the mix of technologies for each year and scenario were used in weighting the production functions for the individual technologies to create a single production function for hydrogen production for each year and scenario.

Estimates were also made of capital expenditures required to build each of the plant types based on the costs of the individual plants and the number of plants built for that year and scenario. These capital costs were added to final demand and consist of estimates of both construction costs by type of construction as well as specific costs for large pieces of equipment. Section A.4.7 provides detailed information on the development and assignment of these costs.

The operating, construction, and capital costs of distributed hydrogen production facilities located at service stations are included in this sector rather than the service station industry, despite their location adjacent to the service station. Similarly, this sector includes all of the additional costs of operating and constructing facilities related to the hydrogen wholesaling, transportation and retail sectors.

Levels of final demand for hydrogen are increased as hydrogen technologies replace gasoline vehicles. Estimates of hydrogen purchases were estimated for each year and scenario. Values were estimated for taxes, hydrogen producer value (at the plant gate), and margins for wholesale, transportation, and retail (service stations). The producer value of hydrogen for each year and scenario was used to increase final demand from hydrogen producers in the Industrial Gas sector 148. Increases in taxes and margins for wholesale, transportation, and retail were added to the appropriate sectors. A key tax assumption was that state and federal excise taxes would not change with the hydrogen scenarios. It was assumed that the tax rates would need to be adjusted to reflect the hydrogen consumption per mile and result in the same annual tax revenges. State sales taxes were applied in the same manner as they are applied to gasoline, assumed to be proportional to fuel sales revenue.

A.4.6.4. Gasoline Stations (IMPLAN Sector #407)

The production function (input vector) for this industry was not altered to reflect changed operating conditions for dispensing hydrogen. The only major differences in the operating costs for a hydrogen station versus a gasoline station are increased energy and maintenance requirements. These additional costs varied by the type of technology and are included in the vectors for the production of hydrogen.

Estimates were also made of the incremental capital expenditures required to build hydrogen service stations. Estimates of costs per station and the number of stations were required. In addition, the capital costs avoided by not providing gasoline were also estimated. The majority of these avoided costs savings are from the elimination of tanks to hold the gasoline. These capital costs consist of estimates of both general construction costs as well as specific costs for large pieces of equipment such as compressors and were added or subtracted, as appropriate, from the investment column in final demand. Section A.4.7 provides detailed information on the development and assignment of these costs.

As noted above, the construction and capital costs of distributed hydrogen production facilities located at service stations are not included in the service station industry, despite their location adjacent to the service station. Instead, these operations are included in the hydrogen production industry.

Final demand for gasoline stations is estimated as the net changes in margins associated with hydrogen and refinery production. These changes vary by scenario and year.

A.4.6.5. Petroleum Refineries (IMPLAN Sector #142)

The production function (input vector) for this industry was not altered to reflect changes wrought by reduction in demand for refined petroleum products.

However, it was assumed that there will be a reduction in capital expenditures as a result of the reduction in demand for refined petroleum products. These avoided costs, which consist of estimates of both general construction costs as well as specific costs for large pieces of equipment represent new facilities or equipment that will no longer be required. These avoided costs were subtracted from final demand Section A.4.7 provides detailed information on the development and assignment of these costs.

Levels of final demand for petroleum products will also be reduced as hydrogen technologies replace internal combustion engine vehicles. Gasoline purchases were estimated for each year and scenario. Demands, which were originally in purchasers' values, were divided among taxes, refinery producer value (at the plant gate), and margins for wholesale, transportation, and retail (service stations). The reduction in producer value from the base case to the scenario for each year was used to reduce final demand from petroleum refineries. Decreases in taxes and margins for wholesale, transportation, and retail were subtracted from the appropriate sectors.

A.4.6.6. Power Generation and Supply (IMPLAN Sector #30)

The production function (input vector) for this industry was not altered to reflect the introduction of stationary fuel cells.[3] However, the demand for electricity of this sector was reduced by the amount of power generated by stationary fuel cells. The demand for natural gas to power the stationary fuel cells was added to final demand; in actuality, these fuel cells would be deployed in the commercial and industrial sectors as well, but the allocation of those uses across a wide array of IMPLAN service sectors that conduct activities in office buildings as well as other industrial sectors was considered impractical. Assignment implicitly to residential use was considered an acceptable simplification in light of the relative magnitudes involved.

Estimates were made of capital expenditures required to purchase the stationary fuel cells. It was also necessary to subtract the cost of avoided power plants construction obviated by the stationary fuel cells. Section A.4.7 provides detailed information on the development and assignment of these costs. These capital costs consist of estimates of both general construction costs as well as specific costs for large pieces of equipment such as turbines and generators in electric utilities and were added to final demand.

A.4.6.7. Wholesale Trade (IMPLAN Sector #391)

The production function (input vector) for this industry was not altered to reflect changes wrought by the shift from petroleum products to hydrogen. The wholesaling of hydrogen will require additional facilities; however, these additional costs are included in the vectors for the production of hydrogen.

Estimates were also made of capital expenditures required to wholesale hydrogen. Capital expenditures were primarily for terminals with liquefiers (trucks) and compressors (pipelines and trucks). The incremental capital costs were included in the hydrogen production sectors and were added to final demand.

Final demands for wholesale services were estimated as the net changes in margins associated with the hydrogen and refinery production. These changes vary by scenario and year.

A.4.6.8. Truck Transportation (IMPLAN Sector #394)

The production function (input vector) for this industry was not altered to reflect changes wrought by the shift from the transportation of petroleum products to hydrogen. It is possible that operation of hydrogen trucks, especially cryogenic trucks, would have a small overall effect on energy inputs.[4]

Estimates were also made of capital expenditures required to deliver hydrogen. Capital expenditures are assumed to be for new vehicles. This required assumptions about the number and costs of vehicles designed to transport hydrogen versus the number and costs of vehicles designed to transport gasoline. The incremental capital costs were estimated as part of the hydrogen production sector capital costs and were added to final demand.

Final demand for truck transportation services are estimated as the net changes in margins associated with the delivery of hydrogen and refinery production. These changes vary by scenario and year.

A.4.6.9. Pipeline Transportation (IMPLAN Sector #396)

The production function (input vector) for this industry was not altered to reflect changes wrought by the shift from the transportation of petroleum products to hydrogen. The operation of hydrogen pipelines will have an overall effect on energy inputs; however, these additional costs are included in the vectors for the production of hydrogen.

Estimates were made of pipeline capital expenditures required to transport hydrogen. Capital expenditures were assumed to be for new pipe required, due to the inability to transport hydrogen in the existing system. The incremental capital costs were added to final demand. Section A.4.7 provides detailed information on the development and assignment of these costs.

Final demand for pipeline transportation services are estimated as the net changes in margins associated with the delivery of hydrogen and refinery production. These changes vary by scenario and year.

A.4.6.10. Other Industries

Additional changes to final demand were made to reflect changing investments in oil and gas well drilling, coal mining, and other sectors. Some of these investments add to final demand while others avoid construction and investment costs that reduce final demand. Section A.4.7 provides detailed information on the development and assignment of these costs.

A.4.7. DEVELOPMENT OF HYDROGEN TECHNOLOGY PRODUCTION FUNCTION VECTORS[5]

The expansion of a hydrogen economy involves the introduction of technologies not in commercial use in today's economy and the deployment of new products and infrastructure as well. In each instance, relevant production functions must be modified to reflect the new technologies in order to gauge the effects of their introduction on employment, both in the industry undergoing change and in the industries supplying it.

The process for accomplishing these modifications involves an examination of the current production function for the target industry and the careful modification of commodity coefficients to reflect the predicted changes in technology predicted. In IMPLAN terms, the production function corresponds to a given industry's column of the gross absorption matrix. It is called "gross" because it represents total commodity needs regardless whether the good is locally purchased or imported. Each term or element in the column or vector represents the industry purchases from another industry.

The introduction of hydrogen technologies involves many changes in the U.S. national economy and in the various regional sub-economies; however, there are two principal changes that must be modeled through modifications to production functions. These two primary changes involve (1) the shift of motor vehicles from today's conventional internal combustion engines to hydrogen fuel cells and (2) the shift of fuel production from gasoline to hydrogen. These key technology changes and their integration into industry production functions are examined in turn below.

A.4.7.1. Hydrogen-Powered Motor Vehicles

Motor vehicles powered by hydrogen fuel cells will closely resemble today's conventional vehicles in all respects except the engine. That is, the body, chassis, suspension, transmission, accessories, etc. will not have to undergo radical change. This group of components is termed the *glider,* and for purposes of the production function, it is assumed to remain unchanged.

The conventional internal combustion engine will be replaced by a hydrogen fuel cell system that will power an electric motor similar to the type utilized today in hybrid vehicles. In addition to the fuel cell, the hydrogen vehicle will require a carbon fiber fuel storage system and a larger and more powerful storage battery. Not only will these changes mean major technological changes in the automobile manufacturing industry, they will also induce even larger changes in the automobile parts industry. As currently defined, the automobile manufacturing industry assembles vehicles, but fabricates few of the components it uses in the assembly process. These components, instead, are fabricated to specification and then purchased from a multitude of parts manufacturers.

In I/O terms, this means that the production functions of upstream and downstream manufacturers must be adjusted. The downstream adjustments to the automobile manufacturing industry (NAICS 344) are the simpler ones. The adjustments in this case affect only three industries that sell supplies or components directly to the assemblers and will be affected directly by the expansion of hydrogen. These industries are shown in Table A.4.8 below. The figures in Table A.4.8 represent the net changes in cost per vehicle realized by the manufacturers due to the conversion to hydrogen as calculated in the technical assessment portion of this project.

For computational ease, it was decided to calculate modifications on a per vehicle basis, and because of the large cost differentials between standard passenger vehicles and sport utility vehicles, it was further decided to differentiate between these two vehicle types. The actual production function coefficients are then calculated based on numbers of each type of vehicle produced in target years.

Table A.4.8. Modifications to Production Function, Motor Vehicle Manufacturing, IMPLAN Sector 344 [in 2005 dollars]

Industry	Dollar Change per Hydrogen Vehicle Produced	
	Mid-sized Passenger Car	Sport Utility Vehicle
Rubber and plastics hose and belting		
180 manufacturing	$13	$22
337 Storage battery manufacturing	$1,681	$2,318
Motor vehicle parts		
350 manufacturing	$2,921	$4,489
Estimated		

Table A.4.9. Modifications to Production Function, Motor Vehicle Parts Manufacturing, IMPLAN Sector 350 [in 2005 dollars]

Industry	Dollar Change per Hydrogen Vehicle Produced	
	Mid-sized Passenger Car	Sport Utility Vehicle
23 Gold; silver; and other metal ore mining	$477	$835
221 Ferrous metal foundries	($946)	($1,368)
222 Aluminum foundries	($872)	($1,262)
342 Carbon and graphite product manufacturing	$775	$1,141
350 Motor vehicle parts manufacturing	$1,471	$1,957
Estimated		

Table A.4.9 shows the impacts realized by the motor vehicle parts manufacturers on purchases from their suppliers. Foundries undergo negative changes in purchases due to the elimination of conventional engine blocks and other metallic products. There are significant increases in purchases for platinum catalysts and for carbon graphite components, particularly fuel storage tanks. A large amount of cost in this industry is represented by intra-industry transfers – one parts firm selling to another – and a significant portion of total cost changes are allocated in this manner. It should also be noted that cost changes realized by the upstream parts industry are less than those of the downstream assemblers. This is because of value-added markups by the parts industry when it sells its products.

A.4.7.2. Hydrogen Production Industries

Hydrogen can be produced from a number of sources. Each technology – coal, nuclear, biomass, wind, and natural gas – represents a unique I/O production function. These technologies will likely be used in combination in the future, thus presenting a challenge to the use of IMPLAN. Thus, for computational purposes, it was decided to develop "pure" production vectors for each technology and combine these vectors according to the appropriate mix under each scenario.

For hydrogen production, Industry 148 Industrial Gas Manufacturing was chosen as the base industry. Hydrogen gas is currently produced commercially in this industry, along with many other gases. It represents a reasonable starting point for production function development. In this case, the production function was modified in absolute dollars, as opposed to the per unit format of the motor vehicle sector. Dollar costs developed for each technology were allocated to appropriate industries and ultimately merged with the base vector quantities. The incremental costs for the centralized production technologies in millions of dollars per year are shown in Table A.4.10 below. (Forecourt production technologies are not shown).

When using absolute figures, it is important to scale the numbers properly. This is a two step process. First, total costs sized to the 2050 economy were scaled to 2003 dimensions using relative real GDP figures. Second, through the use of relative industry revenue figures, Industry 148's coefficients were scaled to the size of the 2003 economy as if fully utilizing hydrogen. These adjustments complete the modifications to Industry 148 needed to represent hydrogen economy production functions. Industries in red font were added to the base vector.

Table A.4.10. Modifications to Production Function, Industrial Gas Manufacturing, IMPLAN Sector 148 [in Millions of 2005 Dollars]

Sector 148 Industrial Gas Manufacturing	**Natural Gas N. America NG SR Central Plant**	**Coal Gasific.+ C Seq. Central Plant**	**Biomass Energy Crop Gasification Central Plant**	**Nuclear Electrolysis Central Plant**	**Nuclear Thermo-chem Central Plant**	**Wind Class 6 Electrolysis Central Plant**
Industry						
Grain farming	$0.00	$0.00	$4,328.89	$0.00	$0.00	$0.00
Sugar cane and sugar beet farming	$0.00	$0.00	$4,328.89	$0.00	$0.00	$0.00
Oil and gas extraction	$2.06	$3.09	$4.12	$5.15	$6.18	$7.21
Coal Mining	$0.00	$5,441.67	$0.00	$0.00	$0.00	$0.00
Power generation and supply	$5,309.27	$4,754.84	$4,169.31	$3,299.84	$4,878.48	$3,299.84
Natural gas distribution	$23,386.95	$0.00	$1,037.91	$0.00	$0.00	$0.00
Water and sewage	$322.16	$97.28	$42.28	$283.36	$226.63	$94.07
Maint. and repair of nonresidential build	$430.21	$640.67	$459.12	$1,488.13	$1,544.22	$1,896.14
Petroleum refineries	$1,056.47	$1,077.61	$1,056.47	$1,056.47	$1,072.26	$1,056.47
Other inorganic chemical manufacture	$713.78	$0.00	$2,549.55	$6,837.24	$6,404.53	$0.00
Machine shops	$172.08	$256.26	$183.65	$595.25	$617.69	$758.45
Metal valve manufacturing	$172.08	$256.26	$183.65	$595.25	$617.69	$758.45
Power boiler and heat exchanger mfg.	$0.00	$0.00	$163.45	$0.00	$0.00	$1,383.79

Table A.4.10. (Continued)

Sector 148 Industrial Gas Manufacturing	**Natural Gas N. America NG SR Central Plant**	**Coal Gasific.+ C Seq. Central Plant**	**Biomass Energy Crop Gasification Central Plant**	**Nuclear Electrolysis Central Plant**	**Nuclear Thermo-chem Central Plant**	**Wind Class 6 Electrolysis Central Plant**
Turbine and Turbine Generator mfr	$0.00	$0.00	$0.00	$0.00	$0.00	$1,476.64
Measuring and dispensing pump mfg.	$169.37	$172.76	$169.37	$169.37	$169.37	$169.37
Industry						
Misc. fabricated metal products mfg.	$72.87	$74.32	$236.32	$1,195.34	$523.81	$1,456.67
Semiconductors manufacturing	$172.08	$256.26	$183.65	$595.25	$617.69	$758.45
All other electronic manufacturing	$172.08	$256.26	$183.65	$595.25	$617.69	$758.45
Miscellaneous electrical equipment	$0.00	$0.00	$0.00	$1,008.23	$450.93	$0.00
Wholesale trade	$172.08	$256.26	$183.65	$595.25	$617.69	$758.45
Rail transportation	$0.00	$604.63	$480.99	$0.00	$0.00	$0.00
Truck transportation	$117.39	$119.74	$117.39	$459.25	$439.37	$117.39
Pipeline transportation	$2,598.55	$0.00	$115.32	$0.00	$0.00	$0.00
Scenic and sightseeing trans. and supply	$46.95	$244.79	$181.97	$299.71	$244.77	$217.62
Warehousing and storage	$46.95	$244.79	$181.97	$299.71	$244.77	$217.62
Telecommunications	$46.95	$244.79	$181.97	$299.71	$244.77	$217.62
Data processing services	$46.95	$244.79	$181.97	$299.71	$244.77	$217.62
Nondepository credit intermediation	$46.95	$244.79	$181.97	$299.71	$244.77	$217.62

Table A.4.10. (Continued)

Sector 148 Industrial Gas Manufacturing	**Natural Gas N. America NG SR Central Plant**	**Coal Gasific.+ C Seq. Central Plant**	**Biomass Energy Crop Gasification Central Plant**	**Nuclear Electrolysis Central Plant**	**Nuclear Thermo-chem Central Plant**	**Wind Class 6 Electrolysis Central Plant**
Industry						
Securities; commodity contracts; invest.	$46.95	$244.79	$181.97	$299.71	$244.77	$217.62
Insurance carriers	$1,259.05	$2,295.61	$1,522.20	$2,723.00	$2,395.70	$3,998.72
Monetary authorities and depository credit	$46.95	$244.79	$181.97	$299.71	$244.77	$217.62
Real estate	$46.95	$244.79	$181.97	$299.71	$244.77	$217.62
Lessors of nonfinancial intangible assets	$46.95	$244.79	$181.97	$299.71	$244.77	$217.62
Legal services	$46.95	$244.79	$181.97	$299.71	$244.77	$217.62
Accounting and bookkeeping services	$46.95	$244.79	$181.97	$299.71	$244.77	$217.62
Architectural and engineering services	$46.95	$244.79	$181.97	$299.71	$244.77	$217.62
Management consulting services	$1,030.00	$11.90	$36.05	$183.11	$168.55	$68.67
Scientific R&D services	$46.95	$244.79	$181.97	$299.71	$244.77	$217.62
All other miscellaneous prof. and tech	$46.95	$244.79	$181.97	$299.71	$244.77	$217.62
Management of companies and enterprises	$46.95	$244.79	$181.97	$299.71	$244.77	$217.62
Waste Mgt. and Remediation Services	$0.00	$466.01	$690.13	$0.00	$0.00	$0.00
Food services and drinking places	$46.95	$244.79	$181.97	$299.71	$244.77	$217.62
Automotive repair and maintenance	$46.95	$244.79	$181.97	$299.71	$244.77	$217.62
Electronic equipment repair and maint.	$215.11	$320.33	$229.56	$744.06	$772.11	$948.06

Table A.4.10. (Continued)

Sector 148 Industrial Gas Manufacturing	Natural Gas N. America NG SR Central Plant	Coal Gasific.+ C Seq. Central Plant	Biomass Energy Crop Gasification Central Plant	Nuclear Electrolysis Central Plant	Nuclear Thermo-chem Central Plant	Wind Class 6 Electrolysis Central Plant
Industry						
Commercial machinery repair and maint.	$215.11	$320.33	$229.56	$744.06	$772.11	$948.06
Federal electric utilities	$187.39	$167.82	$147.16	$340.32	$172.19	$116.46
State and local gov't electric utilities	$749.54	$671.27	$588.60	$465.86	$688.73	$465.86
Other State and local gov't enterprises	$46.95	$244.79	$181.97	$299.71	$244.77	$217.62

[a] An indirect heat gasification process, chosen for the H2A advanced biomass gasification case, is used in this analysis. Estimated

A.4.8. Development of Investment Estimates

A summary view of the investment costs is presented in Tables A.4.11 and A.4.12, at the end of this sub-section.

A.4.8.1. Hydrogen Production and Delivery Infrastructure

Investment costs for the complete hydrogen production and delivery, including dispensing, infrastructure under the HFI Scenario were estimated from Appendix 2 as $595 billion. This cost is consistent with DOE's cost goals. Investments were assumed to be spread equally over a 35-year period, beginning in 2015, resulting in a $17 billion average annual investment. The equal investment magnitudes were assumed for all the sectors for which changes in investment were estimated to avoid creating a sense of greater precision than was possible. From the cost data in Appendix 2, an average of 21% of investment costs were estimated to be construction costs with the remainder being allocated to purchases of equipment whose costs are also identified in Appendix 2. Of the construction costs, 80% were assumed to go to industrial buildings construction and 20% to gas pipeline construction. The distribution of equipment costs across each production technology was calculated; those costs were weighted by each technology's contribution to hydrogen supply and summed to obtain total expenditures on each type of equipment in each of the three scenario years and for both scenarios. The purchases were assigned to IMPLAN sectors that produce the corresponding types of equipment. Investment costs in the Less Aggressive Scenario were estimated to be 2/3 the level of those in the HFI Scenario.

A.4.8.2. Gasoline Stations

Market expansion of hydrogen would permit energy companies to avoid construction costs of for new gasoline stations. In the past decade, the gasoline station infrastructure has changed from a larger number of smaller stations to a smaller number of larger ones. Under the HFI Scenario, it was assumed that construction of 125,000 of the larger gasoline stations could be avoided over the 35-year period covered. An investment cost of $300 thousand per station was assumed, for a total, 35-year cost savings of $37.5 billion, or $1.071 billion per year.[6] Investment costs avoided in the Less Aggressive Scenario were estimated to be 2/3 the level of those in the HFI Scenario.

A.4.8.3. Petroleum Refineries

Industry experts indicated that over the 35-year period in question, the U.S. refining industry would have to build between 10 and 20 new, integrated refineries to meet future gasoline and other refined product demands, at an average cost of $5 billion each.[7] It was assumed that construction of 15 refineries could be avoided under the HFI Scenarios, for a total cost savings of $75 billion, annualized to $2.143 billion. Investment costs avoided in the Less Aggressive scenario were estimated to be 2/3 the level of those in the HFI Scenario.

A.4.8.4. Vehicle and Vehicle Components Factories

It was estimated that the automotive industry would have to invest in construction or modification of 100 plants over the 35-year period, at an average cost of $1 billion each, or a total of $100 billion, annualized to $2.857 billion.[8] There were 56 assembly plants in the United States in the mid-1990s, and nearly twice that number again of components plants owned just by Ford, GM, DaimlerChrysler, and Honda.[9] The estimate of 100 modifications or new constructions relies on such a base of existing facilities and the anticipated time path of the production of hydrogen vehicles under the HFI Scenario. These investments were assumed to be exclusively brownfield construction or modifications of existing plants. Investment costs in the Less Aggressive Scenario were estimated to be 2/3 the level of those in the HFI Scenario.

A.4.8.5. The Oil and Gas Sector

The hydrogen transformation was assumed to permit investments to be avoided in drilling of oil wells and in support activities for oil operations. The current level of activity of $27.4 billion in those activities was assumed to be reduced by 40% under the HFI Scenario, thereby avoiding $11.1 billion per year, distributed roughly 70% to drilling and 30% to support activities.[10] Drilling is associated with new wells, whereas support activities also cover existing wells.

Investment costs avoided in the Less Aggressive Scenario were estimated to be 2/3 the level of those in the HFI Scenario.

A.4.8.6. Coal Mining

Use of coal gasification for roughly half of hydrogen production in 2035 and 2050, in both scenarios, will increase coal output considerably. Estimates are for a 25% increase in coal output in 2035 (coal gasification is not used in 2020 in either scenario) and a 33% increase in 2050 in the HFI Scenario. These increases amount to $5.8 million in 2035 and $15.5 million in 2050, relatively modest magnitudes compared to the other investment changes.[11] Investment costs in the Less Aggressive Scenario were estimated to be 2/3 the level of those in the HFI Scenario.

A.4.8.7. Stationary Fuel Cells

Initially, stationary fuel cells provide back-up power and power for remote locations not easily served by the grid. Users both on the grid and in remote markets are willing to pay a premium for secure reliable electricity. As a backup power source, fuel cells are particularly attractive for hospitals, hotels, data centers, and computer facilities where uninterrupted power is critical. While these initial markets provide the sales volumes needed to launch the fuel cell industry, the fuel cells themselves operate relatively few hours per year and displace little grid electricity. Not until fuel cells begin to penetrate the broad residential and commercial power market does electricity displacement become substantial.

Even though the specific performance requirements for stationary applications differ from transportation applications, some of the technical challenges are the same. For example, the overall cost of stationary fuel cell power systems must be competitive with conventional technologies. Stationary systems, however, have an acceptable price point considerably higher than transportation systems; stationary systems are projected to cost as much as $1000/kW initially but to decline to $400–$750/kW for eventual widespread commercialization.[12]

Like fuel cells for mobile applications, stationary fuel cells are assumed to meet program cost and performance targets. Stationary fuel cells become commercial in 2018 under the HFI Scenario, achieving 1% penetration of "new" electric demand in 2020, 5% in 2035, and 10% in 2050. "New" demand is defined as the difference between electricity demand in 2015 and the analysis year (2020, 2035 or 2050). Applying these penetrations to the demand forecast in this scenario, stationary fuel cells displace 0.01 quad of grid electricity in 2020, 0.3 quads in 2035, and over 1 quad in 2050. Assuming 70% capacity utilization of the fuel cells, these investment costs are estimated at $110 million in 2020, $670 million in 2035, and $3.3 billion in 2050.

Under the Less Aggressive Scenario, stationary fuel cells are also assumed to first penetrate remote and backup power markets, followed by the broad market for residential and commercial electricity, with large-scale deployment beginning in 2020 when program targets are assumed to be met. Stationary fuel cells then achieve 2% penetration of new electric demand in 2035 and 5% in 2050. Thus, they displace approximately 0.1 quad of electricity in 2035 and 0.5 quads in 2050, with corresponding investment costs of $270 million in 2035 and $1.67 billion in 2050.

A.4.8.8. Avoided Utility Investment

It was assumed that avoided utility investments allowed by the investment in stationary fuel cells were roughly equal to the fuel cell investments.

A.4.8.9. Overall Investment

Total nonresidential private fixed investment in the United States in 2005 was $1.446 trillion.[13] Assuming this investment grows at the rate of GDP growth assumed in the analysis above (2.48% per year), the annual investment in hydrogen infrastructure would be 0.84% of total investment in 2020 and 0.58% in 2035 under the HFI Scenario. A total investment cost of $1.5 trillion, instead of $595 billion would increase that share to 3%. Taking into consideration the extent of avoided expenditures permitted by hydrogen, the net annual investment change would be 0.27% in 2020, 0.19% in 2035, and 0.13% in 2050.

Table A.4.11. Changes in Investment Vector, by IMPLAN Sector [in millions of 2005 dollars] HFI SCENARIO

Year	Other new construction (IMPLAN 41)	Turbine and turbine generator set units manufacturing (IMPLAN 285)	Electric power and specialty transformer manufacturing (IMPLAN 333)	Storage battery manufacturing (IMPLAN 337)	Drilling Oil and Gas Wells (IMPLAN 27)	Support activities for oil and gas activities (IMPLAN 28)	Manufacturing and industrial buildings (IMPLAN 37)
Stationary Fuel Cells							
2020	$22	$0	$0	$110	$0	$0	$0
2035	$134	$0	$0	$670	$0	$0	$0
2050	$660	$0	$0	$3,300	$0	$0	$0
Avoided Utility Investment							
2020	($22)	($55)	($55)	$0	$0	$0	$0
2035	($134)	($335)	($335)	$0	$0	$0	$0
2050	($660)	($1,650)	($1,650)	$0	$0	$0	$0
Hydrogen production & delivery infrastructure							
2020	$1,178	$1,488	$0	$0	$0	$0	$2,392
2035	$1,178	$2,988	$0	$0	$0	$0	$2,392
2050	$1,178	$3,089	$0	$0	$0	$0	$2,392
Gas station investments avoided							
2020	($45)	($94)	$0	$0	$0	$0	($180)
2035	($45)	($188)	$0	$0	$0	$0	($180)
2050	($45)	($195)	$0	$0	$0	$0	($180)
Refinery investments avoided							
2020	($90)	($188)	$0	$0	$0	$0	($360)
2035	($90)	($377)	$0	$0	$0	$0	($360)
2050	($90)	($389)	$0	$0	$0	$0	($360)
2020	$0	$250	$0	$0	$0	$0	$600
2035	$0	$502	$0	$0	$0	$0	$600
2050	$0	$519	$0	$0	$0	$0	$600
Oil & Gas exploration avoided							
2020	$0	$0	$0	$0	($3,800)	($3,500)	$0

Year	Other new construction (IMPLAN 41)	Turbine and turbine generator set units manufacturing (IMPLAN 285)	Electric power and specialty transformer manufacturing (IMPLAN 333)	Storage battery manufacturing (IMPLAN 337)	Drilling Oil and Gas Wells (IMPLAN 27)	Support activities for oil and gas activities (IMPLAN 28)	Manufacturing and industrial buildings (IMPLAN 37)
2035	$0	$0	$0	$0	($3,800)	($3,500)	$0
2050	$0	$0	$0	$0	($3,800)	($3,500)	$0
Coal mining investments							
2020	$0	$0	$0	$0	$0	$0	$0
2035	$0	$0	$0	$0	$0	$0	$0
2050	$0	$0	$0	$0	$0	$0	$0
Totals							
2020	**$1,043**	**$1,402**	**($55)**	**$110**	**($3,800)**	**($3,500)**	**$2,452**
2035	**$1,043**	**$2,591**	**($335)**	**$670**	**($3,800)**	**($3,500)**	**$2,452**
2050	**$1,043**	**$1,374**	**($1,650)**	**$3,300**	**($3,800)**	**($3,500)**	**$2,452**
Stationary Fuel Cells							
2020	$0	$0	$0	$0	$0	$0	$0
2035	$0	$0	$0	$0	$0	$0	$0
2050	$0	$0	$0	$0	$0	$0	$0
Avoided Utility Investment							
2020	$0	$0	$0	$0	$0	$0	$0
2035	$0	$0	$0	$0	$0	$0	$0
2050	$0	$0	$0	$0	$0	$0	$0
Hydrogen production & delivery infrastructure							
2020	$4,943	$1,214	$0	$0	$2,650	$2,736	$358
2035	$4,197	$331	$0	$0	$2,058	$2,975	$817
2050	$4,004	$310	$0	$0	$2,078	$3,101	$790
Gas station investments avoided							
2020	($312)	($77)	$0	$0	($167)	($172)	($23)
2035	($264)	($21)	$0	$0	($130)	($187)	($52)
2050	($252)	($20)	$0	$0	($131)	($195)	($50)

Table A.4.11. (Continued)

Year	**Other new construction (IMPLAN 41)**	**Turbine and turbine generator set units manufacturing (IMPLAN 285)**	**Electric power and specialty transformer manufacturing (IMPLAN 333)**	**Storage battery manufacturing (IMPLAN 337)**	**Drilling Oil and Gas Wells (IMPLAN 27)**	**Support activities for oil and gas activities (IMPLAN 28)**	**Manufacturing and industrial buildings (IMPLAN 37)**
Refinery investments avoided							
2020	($623)	($153)	$0	$0	($334)	($345)	($45)
2035	($529)	($42)	$0	$0	($259)	($375)	($103)
2050	($505)	($39)	$0	$0	($262)	($391)	($100)
Auto Plants							
2020	$831	$204	$0	$0	$445	$460	$60
2035	$705	$56	$0	$0	$346	$500	$137
2050	$673	$52	$0	$0	$349	$521	$133
Oil & Gas exploration avoided							
2020	$0	$0	$0	($3,800)	$0	$0	$0
2035	$0	$0	$0	($3,800)	$0	$0	$0
2050	$0	$0	$0	($3,800)	$0	$0	$0
Coal mining investments							
2020	$0	$0	$0	$0	$0	$0	$0
2035	$0	$0	$6	$0	$0	$0	$0
2050	$0	$0	$16	$0	$0	$0	$0
Totals							
2020	**$4,839**	**$1,189**	**$0**	**($3,800)**	**$2,594**	**$2,679**	**$351**
2035	**$4,109**	**$324**	**$6**	**($3,800)**	**$2,015**	**$2,912**	**$800**
2050	**$3,920**	**$304**	**$16**	**($3,800)**	**$2,035**	**$3,036**	**$773**

Table A.4.12. Changes in Investment Vector, by IMPLAN Sector [in millions of 2005 dollars] LESS AGGRESSIVE SCENARIO

Year	Other new construction (IMPLAN 41)	Turbine and turbine generator set units manufacturing (IMPLAN 285)	Electric power and specialty transformer manufacturing (IMPLAN 333)	Storage battery manufacturing (IMPLAN 337)	Drilling Oil and Gas Wells (IMPLAN 27)	Support activities for oil and gas activities (IMPLAN 28)	Manufacturing and industrial buildings (IMPLAN 37)
Stationary Fuel Cells							
2020	$0	$0	$0	$0	$0	$0	$0
2035	$54	$0	$0	$270	$0	$0	$0
2050	$332	$0	$0	$1,600	$0	$0	$0
Avoided Utility Investment							
2020	$0	$0	$0	$0	$0	$0	$0
2035	($54)	($135)	($135)	$0	$0	$0	$0
2050	($332)	($830)	($830)	$0	$0	$0	$0
Hydrogen production & delivery infrastructure							
2020	$789	$997	$0	$0	$0	$0	$1,603
2035	$789	$2,002	$0	$0	$0	$0	$1,603
2050	$789	$2,069	$0	$0	$0	$0	$1,603
Gas station investments avoided							
2020	($30)	($63)	$0	$0	$0	$0	($121)
2035	($30)	($126)	$0	$0	$0	$0	($121)
2050	($30)	($130)	$0	$0	$0	$0	($121)
Refinery investments avoided							
2020	($60)	($126)	$0	$0	$0	$0	($241)
2035	($60)	($252)	$0	$0	$0	$0	($241)
2050	($60)	($261)	$0	$0	$0	$0	($241)
Auto Plants							
2020	$0	$168	$0	$0	$0	$0	$402
2035	$0	$337	$0	$0	$0	$0	$402
2050	$0	$348	$0	$0	$0	$0	$402
Oil & Gas exploration avoided							
2020	$0	$0	$0	$0	($2,546)	($2,345)	$0
2035	$0	$0	$0	$0	($2,546)	($2,345)	$0
2050	$0	$0	$0	$0	($2,546)	($2,345)	$0

Year	Other new construction (IMPLAN 41)	Turbine and turbine generator set units manufacturing (IMPLAN 285)	Electric power and specialty transformer manufacturing (IMPLAN 333)	Storage battery manufacturing (IMPLAN 337)	Drilling Oil and Gas Wells (IMPLAN 27)	Support activities for oil and gas activities (IMPLAN 28)	Manufacturing and industrial buildings (IMPLAN 37)
Coal mining investments							
2020	$0	$0	$0	$0	$0	$0	$0
2035	$0	$0	$0	$0	$0	$0	$0
2050	$0	$0	$0	$0	$0	$0	$0
Totals							
2020	**$699**	**$976**	**$0**	**$0**	**($3,800)**	**($2,345)**	**$1,643**
2035	**$699**	**$1,825**	**($135)**	**$270**	**($3,800)**	**($2,345)**	**$1,643**
2050	**$699**	**$1,196**	**($830)**	**$1,660**	**($3,800)**	**($2,345)**	**$1,643**
Stationary Fuel Cells							
2020	$0	$0	$0	$0	$0	$0	$0
2035	$0	$0	$0	$0	$0	$0	$0
2050	$0	$0	$0	$0	$0	$0	$0
Avoided Utility Investment							
2020	$0	$0	$0	$0	$0	$0	$0
2035	$0	$0	$0	$0	$0	$0	$0
2050	$0	$0	$0	$0	$0	$0	$0
Hydrogen production & delivery infrastructure							
2020	$3,312	$814	$0	$0	$1,775	$1,883	$240
2035	$2,812	$222	$0	$0	$1,379	$1,993	$548
2050	$2,683	$208	$0	$0	$1,393	$2.078	$529
Gas station investments avoided							
2020	($209)	($51)	$0	$0	($112)	($116)	($15)
2035	($177)	($14)	$0	$0	($87)	($126)	($35)
2050	($169)	($13)	$0	$0	($88)	($131)	($33)
Refinery investments avoided							
2020	($417)	($103)	$0	$0	($224)	($231)	($30)
2035	($354)	($28)	$0	$0	($174)	($251)	($69)
2050	($338)	($26)	$0	$0	($176)	($262)	($67)

Year	Other new construction (IMPLAN 41)	Turbine and turbine generator set units manufacturing (IMPLAN 285)	Electric power and specialty transformer manufacturing (IMPLAN 333)	Storage battery manufacturing (IMPLAN 337)	Drilling Oil and Gas Wells (IMPLAN 27)	Support activities for oil and gas activities (IMPLAN 28)	Manufacturing and industrial buildings (IMPLAN 37)
Auto Plants							
2020	$557	$137	$0	$0	$298	$308	$40
2035	$473	$37	$0	$0	$232	$335	$92
2050	$451	$35	$0	$0	$234	$349	$89
Oil & Gas exploration avoided							
2020	$0	$0	$0	($2,546)	$0	$0	$0
2035	$0	$0	$0	($2,546)	$0	$0	$0
2050	$0	$0	$0	($2,546)	$0	$0	$0
Coal mining investments							
2020	$0	$0	$0	$0	$0	$0	$0
2035	$0	$0	$4	$0	$0	$0	$0
2050	$0	$0	$9	$0	$0	$0	$0
Totals							
2020	$3,242	$797	$0	($2,546)	$1,738	$1,795	$235
2035	$2,753	$217	$4	($2,546)	$1,350	$1,951	$536
2050	$2,626	$203	$9	($2,546)	$1,363	$2,034	$518

Source: reported in footnotes to Section A.4.8.

End Notes

[1] Minnesota IMPLAN Group, Inc., *IMPLAN Professional, Version 2.0, Social Accounting &Impact Analysis Software, User Guide, Analysis Guide, and Data Guide*, 3rd Edition Stillwater, Minn., February 2004.

[2] Regional Economic Models, Inc., *REMI Policy Insight Version 8.0.9, 70 Sector Model*, Amherst, Mass., March 22, 2006.

[3] Stationary fuel cells are a small portion of the initial market relative to the transportation fuel cell demand.

[4] Stationary fuel cells are a small portion of the initial market relative to the transportation fuel cell demand.

[5] The production function is described in Section A.4.3.1 Overview of the Process, page 126.

[6] Alabama Cooperative Extension Service, Alabama A&M and Auburn Universities, Agricultural and Natural Resources, "Underground Storage Tanks (USTs) and NPS Pollution; Managing Privately Owned USTs: Guidelines for Private Property Owners," ANR-790-4.8.5, June 1995; Mueller, E.A., "Methanol Refueling Station Costs," prepared for American Methanol Institute, February 1999; TIAX (Acurex) experience in installing over 20 fueling stations.

[7] Interviews with refining industry executives.

[8] Accessed August-September 2006: Toyota New Texas Plant ($800 million) at http://www.toyota.com/about/operations/manufacturing/texas/index.html; Toyota New Ontario Plant (C$800 million), at http://www.cbc.ca/money/story/2005/06/30/toyota-050630.html; Toyota North America Operations, at http://www.toyota.com/about/operations/manufacturing/index.html; Nissan North America Operations, at *http://www.nissannews.* com/ corporate/ manufacturing/tnmfgfacts.shtml; Hyundai U.S. Operations, at *http://www.hyundaiusa.* com/ abouthyundai/usinvestment/usinvestment.aspx.

[9] U.S. Environmental Protection Agency, Common Sense Initiative, Automobile Manufacturing Sector, "U.S. Automobile Assembly Plants and their Communities, Environmental,Economic and Demographic Profile, Part II – Industry-Wide Data," December 1997, p. II-1; accessed August-September 2006 at: http://en.wikipedia.org/wiki/List_of_GM_factories; *http://en .wikipedia*.org/wiki/List_of_Ford_factories; http://en.wikipedia.org/wiki/List_of_DaimlerChrysler_factories; http://corporate.honda.com/america/facilities.aspx.

[10] Accessed August 2006 at http://www.census.gov/prod/ec02/ec0221i213112.pdf and *http://www. census.gov/* prod/ec02/ec0221i213111.pdf.

[11] Accessed August 2006 at http://www.census.gov/prod/ec02/ec0231i333131.pdf.

[12] U.S. Department of Energy, Hydrogen, *Fuel Cells & Infrastructure Technologies Program, Multi-Year Research, Development and Demonstration Plan*, Feb. 2005, accessed September 2006 at http://www1.eere.energy.gov/hydrogenandfuelcells/mypp.

[13] Accessed August 2006 at http://www.bea.gov/bea/an/nipaguid.pdf.

APPENDIX 5. METHODOLOGY FOR EMPLOYMENT CREATION AND REPLACEMENT ESTIMATES

Section 3 of the preceding study presented estimates of the effects of a hydrogen transformation on net U.S. employment, as well as more detailed estimates of skill change-induced job creation and replacement in several key hydrogen-related sectors. It also presented an analysis of the challenges of training and retraining the workforce during the market transformation.

This appendix (Appendix 5) describes (a) the calculation of total sector employment gains and losses using IMPLAN projections, (b) the sources and methods used to develop estimates of job creation and replacement at the occupational level induced by skill changes, and (c) the sources of data relevant to retraining needs among engineers and automotive service technicians and mechanics.

A.5.1. CALCULATION OF TOTAL SECTOR LEVEL GAINS AND LOSSES

Table 3.2 reports gains and losses resulting from movement of employment between sectors. These data were obtained as follows:

(1) The IMPLAN model produced estimates of how much larger or smaller employment in each sector would be as a result of a hydrogen transformation. For instance, the model predicts that the hydrogen

production sector would employ about 73,549 more workers in 2050 under the HFI Scenario than it would have in that year given a continuing gasoline economy.

(2) Using a coded spreadsheet, the positive and negative effects on sector employment were calculated and sorted to derive total gains by sectors that grow more important in a hydrogen economy and total losses by sectors that grow less important. These are the numbers listed as gains and losses in Table 3.1, with the net effect being the difference between gains and losses.

(3) The percent effects on total U.S. employment were obtained by dividing the gains, losses, and net effects by IMPLAN's projection of the base case employment in all sectors for 2020, 2035, and 2050 respectively.

A.5.2. Sources of Employment Data at the Occupational Level

Occupational information for each key sector was obtained to estimate the employment creation and replacement effects of expanding hydrogen markets. This was accomplished using both 4-digit and 6-digit North American Industrial Classification System (NAICS) data from the Bureau of Labor Statistics.[1] Occupational information was combined with output from the IMPLAN model to determine employment impacts under the HFI and Less Aggressive Scenarios. In addition, estimates were made of the effects of switching to production in a hydrogen economy on the occupational distributions of key sectors based on the consensus opinions of industry experts. Some of the consensus opinions they expressed on the nature of likely skill changes are summarized in Sections 3.4.1 and 3.4.2 of the report. The most current information from BLS was used, including May 2005 for the 4-digit NAICS and 2004 for the 6-digit NAICS. The combined use of data was necessary because complete data containing specific occupations are only available at the 4-digit NAICS level. Sectors specified by the IMPLAN model are more detailed, and total numbers of employees are specified only at the 6-digit industry level. Table A.5.1 lists each IMPLAN sector and the corresponding 4-digit NAICS category that was used to obtain its occupational distribution data.

Table A.5.1. List of IMPLAN Sectors and Corresponding NAICS Codes and Titles Used to Obtain Occupational Distribution Data

IMPLAN Sector	IMPLAN Description	4-digit NAICS Code	NAICS Description
37	Manufacturing and industrial buildings	2362	Non-residential building construction
40	Water, sewer, and pipeline construction	2371	Oil and gas pipeline construction
142	Petroleum refineries	3241	Petroleum and coal products manufacturing
147	Petrochemical manufacturing	3251	Basic chemical manufacturing
148	Industrial gas manufacturing	3251	Basic chemical manufacturing
344	Automobile and light truck manufacturing	3361	Motor vehicle manufacturing
346	Motor vehicle body manufacturing	3362	Motor vehicle body and trailer manufacturing
350	Motor vehicle parts manufacturing	3363	Motor vehicle parts manufacturing
401	Motor vehicle and parts dealers	441	Automobile dealers (4411) & Automotive parts, accessories, and tire stores (4413)
483	Automotive repair and maintenance, except car washes	8111	Automotive repair and maintenance

Source: Minnesota IMPLAN Group, "2001 IMPLAN Sectoring Scheme – Replacement for Appendix A in IMPLAN User's Manual."

A distribution of occupation percentages was calculated for each sector and was then applied to IMPLAN model results. Estimates of job creation and replacement within key sectors (shown in Tables 3.2 and 3.3 of the report) as well as the average annual percent employment impacts (Tables 3.4 and 3.6) were produced using coded spreadsheets based on the sector level occupational distributions. The calculations carried out in the spreadsheets are described below.

A.5.3. Summary of the Calculation and Consistency Requirements

The overall function of the spreadsheets was to take estimates of the number of workers in each sector that would be affected by skill changes, given a complete transformation of markets to hydrogen and use them to produce estimates of the number of workers affected at different dates in the transformation under different scenarios. The estimates for the three years under each scenario had to correspond with the results of the IMPLAN model, which produced estimates of changing total sector sizes under each scenario. Thus, sector estimates had to be consistent with any overall sector growth or decline.

A.5.4. Raw Creation and Replacement Estimates

The main inputs into each sector spreadsheet were the 2005 occupational breakdowns whose IMPLAN-NAICS correspondences are shown above in Table A.5.1. Based upon opinions from industry experts, occupations likely to be affected by the hydrogen economy were put into 3 groups: skill affected white collar, skill-affected blue collar, and skill-affected non-production. All remaining occupations were categorized as non-skill-affected. The industry experts were asked questions regarding the number and composition of white collar (engineering occupations), and blue collar occupations. Interviewees were asked (1) to describe how employment would be different, and why, in each key sector; (2) how they envision the development of technology; (3) what further implications a switch to hydrogen might have; and (4) what they saw as the biggest impediments to implementing a hydrogen economy.

Three scenarios were applied to simulate high, medium, and low potential employment impacts for the three groups. Employees could lose jobs due to changes in required skills; due to a change in plant, store, or job location; or due to a combination of both. In addition, for selected occupations, changes were made based on future technology impacts (e.g., overall, there might be fewer mechanical engineers and more electrical engineers due to vehicle design).

The initial estimates represent jobs lost and created due to a complete hydrogen market transformation, all other things being equal. These estimates assume no change in the size of the sector due to population change, demand change for the sector's product, productivity changes, or anything else. For that reason, the total number of workers in each sector was held constant when the estimates were made. If employment in one occupation is estimated to grow, employment in another occupation within the sector is expected to fall. For example, an increase in the relative importance of chemical engineers in the auto parts industry would be countered by a fall in the relative importance of mechanical engineers in that industry.

Changes in employment resulting from non-skill-based total sector employment changes, such as due to product demand, are based on IMPLAN estimates that are dealt with in calculations discussed below. The one non-skill-based change included in the raw estimates is that of employment changes resulting from factory relocations. Job loss and creation estimates are included for non-skill-affected workers based on the expectation that some facilities will relocate.

A.5.5. DATA ADJUSTMENT

The IMPLAN model begins in 2005 with the initial total employment data for each sector involved. Unfortunately, the sectors defined in the IMPLAN model sometimes differ in size from their corresponding sectors in the BLS occupational breakdowns. It was necessary to make the number of workers in each occupation add up to the IMPLAN total in order to use IMPLAN data to add sector growth and decline effects into the skill change estimates.

The simplest way to deal with this problem was to take the ratio of total sector employment according to IMPLAN to total sector employment according to BLS and multiply it by the number of workers in each occupation. So, for industrial engineers in auto parts manufacturing:

> (12,822 industrial engineers in IMPLAN parts sector) = (14,460 industrial engineers according to BLS) x [(615,360 total parts sector jobs in IMPLAN) (693,120 total parts sector jobs according to BLS)]

The same process is then repeated for the job loss and creation estimates, so that they remain proportional to the initial number of workers in the occupation.

While using this ratio is the simplest way to make the adjustment, it implicitly assumes that the distribution of occupations across IMPLAN's version of a sector is the same as that of BLS's version. In some cases, a review of BLS's definitions revealed that such an assumption was reasonable. In other cases, certain groups of occupations had to be added or taken out of the BLS group to produce distributions that were more likely to reflect the realities of the sectors as defined by IMPLAN.

For instance, the IMPLAN sector "Automotive repair and maintenance" consists of three 5- to 6-digit sub-occupations: 81111, 811191, 811198. But, the occupational distribution is only available at the 5-digit level for 81111 and 81119. In addition, sector 81119 includes car washes, which represent over 50% of the employees in the 81119 sector, "Other automotive repair and maintenance." In IMPLAN, car washes are specifically excluded. Because the nature of the work in car washes is substantially different that that of automotive repair technicians, the distribution for sector 81119 could not be applied. Therefore, the number of employees from sectors 811191 and 811198 were added to sector 81111, and the distribution was used for all three because it was most representative of the occupational distribution for all employees.

A.5.6. Splitting up Occupations by Skill

The spreadsheet produces estimates of the number of workers in each occupation in the sector in question at the start of the IMPLAN model in 2005, as well as the number of jobs estimated to be lost and created if gasoline is completely replaced by hydrogen. The spreadsheet then adds the jobs created and subtracts those that are lost to produce a list of the number of workers in each occupation in a hydrogen sector. So, for Electrical Engineers in Motor Vehicle Assembly:

(142 engineers in a gasoline economy) – (36 jobs lost in transformation) + (875 jobs created in transformation) = (981 engineers in a hydrogen economy)

As noted above, the net growth in electrical engineers is offset by declines in other occupations, keeping total sector employment constant for the time being.

The problem posed by this calculation is that its results conceal some job loss and creation. Since the spreadsheet's future calculations will depend on the first and last numbers in the equation, that is, the number of engineers in each type of economy, the 36 jobs lost and 36 of the jobs created will cancel each other out, and it will appear as if the only change that occurred was the addition of 780 engineers to the sector. The fact that 36 workers are projected to be replaced is completely missed in later analysis.

To avoid this problem, the spreadsheets divide each skill-affected occupation into three separate categories: (a) workers whose skills, with some retraining, can apply to both internal combustion engine and hydrogen vehicles, (b) workers whose skills apply only to the internal combustion engine vehicles, and (c) workers whose skills apply only to hydrogen vehicles. The first category shall henceforth be referred to as workers with transferable skills.

The calculation shown above is now done separately for each category:

(107 engineers with transferable skills in a gasoline economy) – (0 jobs lost in transformation) + (0 jobs created in transformation) = (107 engineers with transferable skills in a hydrogen economy)

(36 engineers with internal combustion engine-only skills in a gasoline economy) – (36 jobs lost in transformation) + (0 jobs created in transformation) = (0 engineers with internal combustion engine-only skills in a hydrogen economy)

(0 engineers with hydrogen-only skills in a gasoline economy) – (o jobs lost in transformation) + (875 jobs created in transformation) = (875 engineers with hydrogen-only skills in a hydrogen economy)

All changes in skill-affected occupations were calculated in this manner.

A.5.7. CALCULATING PERCENT DISTRIBUTIONS

Using the methodology described above, the spreadsheet produced a skill-based occupational composition for both fully gasoline and fully hydrogen economies in raw numbers of workers, but these numbers were not be adequate for the coming calculations.

Since the transformation of the economy to hydrogen does not occur in a vacuum, the total sector size for the gasoline economy will be different for 2020, 2035, and 2050 than it is in 2005, so the number of jobs in each occupation in a gasoline economy must be recalculated for each year. The same applies to a hydrogen economy. The recalculation is simplest when each occupation is assigned a percent of the sector's employment, in which case the spreadsheet calculates those shares for gasoline and hydrogen economies.

The percentages were obtained by dividing the number of workers in an occupation by the total number of workers in the sector. The electrical engineers discussed above are part of a sector that employs 243,550 workers, so their percent shares are as follows:

Gasoline Economy:

107 engineers with transferable skills comprise 0.04% of the sector
36 engineers with internal combustion engine-only skills comprise 0.01% of the sector
0 engineers with hydrogen-only skills comprise 0% of the sector

Hydrogen Economy:

107 engineers with transferable skills comprise 0.04% of the sector
0 engineers with internal combustion engine-only skills comprise 0% of the sector
875 engineers with hydrogen-only skills comprise 0.36% of the sector

A.5.8. TAKING WEIGHTED AVERAGES

The percent distributions of occupations, as calculated above, represent two absolutes: a 100% gasoline economy and a 100% hydrogen economy. In most of the years for which the spreadsheet produces estimates, a mix of gasoline and hydrogen in the auto industry is calculated, with hydrogen increasing over the years at a brisk pace in the HFI Scenario and at a slower pace in the Less Aggressive Scenario.

The two scenarios provide assumptions about the mix of gasoline and hydrogen in each of the key years. The mix is measured by the share of vehicles sold that are hydrogen powered and by the share of the stock of vehicles on the road that are hydrogen powered. Depending on the products of each key sector, the sector's employment is expected to make a transition at the same rate as vehicle sales or stock. The sectors and their transition rates are shown in Table A.5.2 below:

Table A.5.2. Transition Rates of Skill Changes in Sectors

Sector	Transition Rate
Automobile Assembly	Transitions with Sales
Auto Body Manufacturing	Transitions with Sales
Auto Parts Manufacturing	Transitions with Sales
Auto Dealers and Repair	Mixed, Depends on Occupation
Hydrogen Production	Transitions with Stock
Construction	Transitions with Stock

Source: estimated in this study.

Since neither of the percent distributions calculated above applies to an automobile industry that uses a mix of fuels, calculation of a new distribution was needed for years when there is such a mix. The new distribution was obtained by taking a weighted average of the gasoline and hydrogen distributions.

The percentage of motor vehicle assembly workers who are team assemblers with internal combustion engine-only skills in 2020 under the HFI Scenario is obtained as follows. Since motor vehicle assembly is weighted by sales, and 26.6% of vehicles sold in 2020 will be hydrogen powered under the HFI Scenario, 26.6% of the sector will follow the hydrogen employment distribution and 73.4% of the sector will follow the gasoline distribution. The calculation of the internal combustion engine-only team assemblers' share is shown below:

(3.54% share of a gasoline sector) x (73.4% gasoline in 2020) + (0% share of the hydrogen sector) x (26.6% hydrogen) = (2.60% share of a 2020 HFI sector)

These new shares were calculated for every occupation for each year in each scenario.

A.5.9. CALCULATING NUMBERS OF WORKERS

Once a percent distribution was calculated for a particular year under a particular scenario, it was used to calculate the number of workers in each occupation. Two numbers were calculated: a baseline number representing a continued all-gasoline economy and a transition number for whichever scenario was being used. The numbers were obtained simply by multiplying the relevant percent share of the occupation by the relevant IMPLAN estimate of total sector size. For the internal combustion engine team assemblers, that proceeds as follows:

Baseline employment, 2020:

(3.54% share of gasoline based sector) x (221,804 workers in the sector in 2020 with no transformation) = (7,852 assemblers with internal combustion engine-only skills in the 2020 baseline scenario)

HFI employment, 2020:

(2.6% share of mixed sector) x (221,804 workers in the sector in 2020 with HFI transformation) = (5,767 assemblers with internal combustion engine-only skills in the 2020 HFI scenario)

A.5.10. Job Creation and Replacement by Scenario and Year

Creation and replacement for each occupation were calculated using the newly obtained employment numbers by occupation. Job creation and replacement for hydrogen transformation in 2020 refer to changes from what employment would have been in 2020 under the Baseline Scenario, so the employment change is calculated by subtracting baseline employment from HFI (or, when appropriate, Less Aggressive) Scenario employment. If the difference is positive, it was placed in the creation column. If it is negative, it was placed in the replacement column.

A.5.11. Obtaining Totals for White Collar and Blue Collar Workers

The occupations in each sector are divided into four categories. One of those is the non-skill-affected workers mentioned earlier. The skill-affected workers are divided into a further three categories: white collar, blue collar, and non-production. There are white collar, blue collar and non-production workers among the non-skill-affected workers as well, but the effects on them are different.

White collar jobs are defined as production occupations that require a post-secondary degree, blue collar jobs are defined as production occupations requiring vocational or employer-provided on-the-job training, and non-production jobs are defined as occupations not directly involved in designing, producing or servicing the physical product to be sold.

To obtain final estimates for a particular year and scenario, the spreadsheet sums the employment creation numbers for each category, then the replacement numbers for each category. The resulting numbers represent total creation and replacement of jobs in the skill-affected white collar, skill-affected blue collar, skill-affected non-production and non-skill-affected categories. The numbers are cumulative, meaning they represent all creation and destruction in the category in question from 2018 to the year in question.

The creation and replacement (loss) numbers for skill-affected white collar and skill-affected blue collar jobs are the numbers that appear in Tables 3.2 and 3.3 of the report, with the net numbers equaling creation less replacement.

Tables 3.4 and 3.6 of the report divide these creation and replacement numbers by the base-case employment numbers for each category of job to get the percent creation and replacement resulting from a transformation to hydrogen. These percents are also cumulative. The creation and replacement percentages are then divided by the number of years elapsed since 2018 to get the average annual rates of creation and replacement that appear in the tables.

A.5.12. Sources of Data and Estimates Related to Retraining Analysis

Table 3.5 presents the total number of specific types of engineers working in key sectors, compared with the total number of specific types of engineers in the U.S. working in all sectors, obtained from the Bureau of Labor Statistics' Occupational Employment Statistics, from May 2005.

Table 3.7 presents the potential number of automobile service technicians that would need to be trained as hydrogen fuel cell vehicles are introduced. The 20% estimate for 2020 is based upon information from industry experts and judgment. This type of estimate was needed because only 3% of the vehicle stock[2] will be hydrogen fuel cell vehicles, but there will need to be a more substantial work force in place to service the vehicles when they are delivered to customers. For 2035 and 2050, the number of trained technicians follows the share of hydrogen vehicles in the stock.

End Notes

[1] Bureau of Labor Statistics (BLS), Occupational Employment Statistics, data as of May 2005: http://data.bls.gov/oes/search.jsp?data_tool=OES.

[2] *Hydrogen Posture Plan*, U.S. Department of Energy report, Feb. 2004, accessed August-September 2006 at http://www1.eere.energy.gov/hydrogenandfuelcells/posture_plan04.html.

INDEX

N

O

P

R